AF719749

Compressed Sensing Approach to Systems and Control

Masaaki Nagahara

now
the essence of knowledge

EURASIP–Now Publishers Open Access Book Series on Information and Learning Sciences

Editor-in-chief

Ali H. Sayed *(EPFL, Switzerland)*

Editors

Helmut Bolcskei *(ETH Zurich, Switzerland)*
Alfred O. Hero *(University of Michigan, USA)*
Angelia Nedich *(Arizona State University, USA)*
H. Vincent Poor *(Princeton University, USA)*
Sergios Theodoridis *(University of Athens, Greece)*
Abdelhak Zoubir *(Technical University Darmstadt, Germany)*

Other titles in EURASIP–Now Publishers Open Access Book Series on Information and Learning Sciences

Social Learning: Opinion Formation and Decision-Making over Graphs
Vincenzo Matta, Virginia Bordignon, Ali H. Sayed
ISBN: 978-1-63828-472-7

Published, sold and distributed by:
now Publishers Inc.
PO Box 1024
Hanover, MA 02339
United States
Tel. +1-781-985-4510
www.nowpublishers.com
sales@nowpublishers.com

Outside North America:
now Publishers Inc.
PO Box 179
2600 AD Delft
The Netherlands
Tel. +31-6-51115274

EURASIP–Now Publishers Open Access Book Series on Information and Learning Sciences

ISBN: 978-1-63828-504-5
E-ISBN: 978-1-63828-505-2
DOI: 10.1561/9781638285052

Copyright © 2025 Masaaki Nagahara

Suggested citation: Masaaki Nagahara. (2025). *Compressed Sensing Approach to Systems and Control.* Boston–Delft: Now Publishers

Contents

Preface

Compressed sensing, also known as sparse representation or sparse modeling, has experienced substantial growth in research fields such as signal processing, machine learning, and statistics. In recent years, this powerful tool has been successfully applied to the design of control systems. This book provides a comprehensive guide to compressed sensing-based techniques with their application to systems and control.

This book is intended for graduate students and researchers who already have a foundational understanding of basic calculus and linear algebra. Its primary objective is to equip readers with the practical skills to apply compressed sensing techniques to a range of engineering problems, with a particular emphasis on systems and control. It also presents a comprehensive collection of efficient algorithms for addressing these problems. Moreover, the book includes accompanying Python programs, which enable readers to actively experiment with these algorithms firsthand. By engaging with these practical examples, readers will develop a deeper understanding of compressed sensing techniques and their applications to systems and control. The Python programs can be downloaded from the following web page:
`https://nagahara-masaaki.github.io/spm_en`

This book is the second edition of the author's previous work, *Sparsity Methods for Systems and Control*, published by Now Publishers in 2020. This edition incorporates significant updates to reflect the latest advancements in the field. Notably, it includes new chapters and sections covering the following key topics:

- Distributed optimization (Chapter 6)
- Sparse system identification (Section 7.2)
- Sparse controller design (Section 7.3)
- Distributed hands-off control (Section 11.4)

I trust that this book will serve as a valuable resource for readers seeking to gain a comprehensive understanding of the state-of-the-art in this field.

This book is organized as follows. Chapter 1 provides a brief overview of the history and literature of compressed sensing. The book is then divided into two parts. Part I (Chapters 2–7) offers a comprehensive foundation in compressed sensing within the context of finite-dimensional vector spaces. Part II (Chapters 8–11) introduces maximum hands-off control, an optimal control strategy based on compressed sensing principles, for continuous-time systems.

Part I commences with Chapter 2, which establishes the fundamental concept of sparsity and its significance in compressed sensing. Chapter 3 introduces sparse optimization, incorporating illustrative examples such as curve fitting and group testing. Chapter 4 provides efficient algorithms for convex optimization, a powerful framework for solving sparse optimization problems. Chapter 5 explores greedy algorithms as alternative approaches to compressed sensing. Chapter 6 extends the scope of sparse optimization to distributed optimization scenarios. Chapter 7 showcases applications of compressed sensing within the context of systems and control, demonstrating the practical utility of the techniques presented in Part I.

Part II begins with Chapter 8, which provides a brief review of dynamical systems and optimal control theory. Chapter 9 introduces the novel concept of maximum hands-off control, an optimal control strategy characterized by its sparsity. Chapter 10 presents a numerical optimization method for solving the optimal control problem associated with maximum hands-off control. Finally, Chapter 11 explores advanced topics in maximum hands-off control, including smooth hands-off control, discrete-valued control, and distributed hands-off control.

Acknowledgements

This work was partly supported by JSPS KAKENHI Grant Number 23K26130.

Masaaki Nagahara

Notation

A finite-dimensional vector is represented in a bold face, e.g. $\boldsymbol{x}$, when the size of the vector is greater than 2. For one-dimensional vectors, we do not use a bold face and simply write like x, regarding as a scalar.

We denote by $\mathbb{R}^n$ the set of n-dimensional real column vectors, and by $\mathbb{R}^{m\times n}$ the set of $m \times n$ real matrices. The transpose of a vector $\boldsymbol{x}$ and a matrix A are respectively denoted by $\boldsymbol{x}^\top$ or $A^\top$. The i-th element of a vector $\boldsymbol{x}$ and the (i,j)-th element of a matrix A are respectively denoted by $(\boldsymbol{x})_i$ and $[A]_{ij}$. We denote by $\mathbb{Z}$ the set of integers and by $\mathbb{N}$ the set of natural numbers, that is, $\mathbb{N} = \{1, 2, 3, \ldots\}$.

For a vector $\boldsymbol{x} \in \mathbb{R}^n$, $\mathrm{supp}(\boldsymbol{x})$ denotes the support set of $\boldsymbol{x}$, that is, the set of nonzero elements of $\boldsymbol{x} = [x_1, \ldots, x_n]^\top \in \mathbb{R}^n$:

$$\mathrm{supp}(\boldsymbol{x}) \triangleq \{i \in \{1, \ldots, n\} : x_i \neq 0\}. \tag{1}$$

The ℓ^0 norm of $\boldsymbol{x} \in \mathbb{R}^n$ is defined by

$$\|\boldsymbol{x}\|_0 \triangleq \#(\mathrm{supp}(\boldsymbol{x})), \tag{2}$$

where $\#(\cdot)$ returns the number of elements of the argument set. The ℓ^p norm with $p \geq 1$ is defined by

$$\|\boldsymbol{x}\|_p \triangleq \left\{\sum_{i=1}^{n} |x_i|^p\right\}^{1/p}, \tag{3}$$

and the ℓ^∞ norm by

$$\|\boldsymbol{x}\|_\infty \triangleq \max_{i=1,2,\ldots,n} |x_i|. \tag{4}$$

In Part II, these norms will be denoted by $\|\boldsymbol{x}\|_{\ell^0}$, $\|\boldsymbol{x}\|_{\ell^p}$, and $\|\boldsymbol{x}\|_{\ell^\infty}$ to distinguish them from the norms used for continuous-time signals.

For a vector $\boldsymbol{x} \in \mathbb{R}^n$, and an index set $S \subset \{1, 2, \ldots, n\}$, we denote by $\boldsymbol{x}_S$ the restriction of $\boldsymbol{x}$ to S. Namely, if $\boldsymbol{x} = [x_1, x_2, \ldots, x_n]^\top$ and $S = \{i_1, i_2, \ldots, i_k\}$ $(1 \leq i_1 < i_2 < \cdots < i_k \leq n)$, then

$$\boldsymbol{x}_S = [x_{i_1}, x_{i_2}, \ldots, x_{i_k}]^\top \in \mathbb{R}^k. \tag{5}$$

Also, for $\Phi = [\phi_1, \phi_2, \ldots, \phi_n] \in \mathbb{R}^{m\times n}$, Φ_S is defined as

$$\Phi_S = [\phi_{i_1}, \phi_{i_2}, \ldots, \phi_{i_k}] \in \mathbb{R}^{m\times k}. \tag{6}$$

The complement of a set $\mathcal{S}$ is denoted by $\mathcal{S}^c$.

Let $f : [0,T] \to \mathbb{R}$ be a measurable function with $T > 0$. The support of f is denoted by $\mathrm{supp}(f)$ and defined by

$$\mathrm{supp}(f) \triangleq \{t \in [0,T] : f(t) \neq 0\}. \tag{7}$$

The L^0 norm of f is defined by

$$\|f\|_0 \triangleq \mu(\mathrm{supp}(f)), \tag{8}$$

where μ is the Lebesgue measure over $\mathbb{R}$. The L^p norm with $p \geq 1$ is defined by

$$\|f\|_p \triangleq \left\{ \int_0^T |f(t)|^p dt \right\}^{1/p}, \tag{9}$$

and the L^∞ norm by

$$\|f\|_\infty \triangleq \sup_{t\in[0,T]} |f(t)|. \tag{10}$$

We denote by $L^p(0,T)$ with $p \geq 1$ or $p = \infty$ the set of functions with finite L^p norm.

For a function $f : \mathbb{R}^n \to \mathbb{R}$, the *gradient* ∇f is defined by

$$\nabla f \triangleq \frac{\partial f}{\partial \boldsymbol{x}} = \left[\frac{\partial f}{x_1}, \frac{\partial f}{x_2}, \ldots, \frac{\partial f}{x_n}\right]^\top \in \mathbb{R}^n. \tag{11}$$

We say a real-valued function $f(n)$, $n \in \mathbb{N}$, is $O(g(n))$ if

$$\limsup_{n\to\infty} \left|\frac{f(n)}{g(n)}\right| < \infty.$$

Chapter 1

Introduction

In this chapter, we briefly review the history of compressed sensing, also known as sparse modeling, in science and engineering. The chapter will motivate you to learn this topic. The content of this chapter is independent of the other chapters, and readers interested in the technical aspects of compressed sensing can skip this chapter without much effect on their understanding of the rest of the book.

1.1 Occam's Razor

At the root of sparsity methods, including compressed sensing, lies the idea that one should not assume more than is necessary to explain certain things. This is known as *Occam's razor*, also called the *law of parsimony*, developed by Ockham in the 14th century. This idea was not invented by Ockham but rather long before him, for example, by Claudius Ptolemy (90AD–168AD) and Aristotle (384BC–322BC). This is a very familiar concept to us, especially in Japan, where there is a culture of Zen and Wabi/Sabi, which can be roughly translated as *"simple is best."*

There is a satirical depiction of the opposite of Occam's razor, *Rube Goldberg machine.* Figure 1.1 shows an example of a Goldberg machine. The machine is a self-operating napkin, which automatically wipes off the dirt from the beard when he drinks soup. This caricature depicts a machine that performs very simple actions with extreme complexity and satirizes the large-scale mechanization in the first half of the twentieth century. The machine runs as follows.

1. The man raises the spoon of soup (A) to his mouth.
2. The string (B) attached to the spoon (A) is pulled.

Figure 1.1: Rube Goldberg machine (self-operating napkin)

3. The ladle (C) moves.
4. The cracker (D) flies on the parrot (E).
5. The parrot (E) takes off after the cracker (D).
6. The perch (F) tilts.
7. The seeds (G) on the perch (F) spill out and go into the pail (H).
8. The string (I) is pulled by the extra weight in the pail (H).
9. It ignites the cigar lighter (J).
10. The fuse of the rocket (K) is lit and it takes off.
11. The knife (L) attached to the rocket (K) cuts the string (M).
12. The pendulum swings and the napkin (N) wipes the dirt off the beard.

The Goldberg machine is obviously strange. However, in this highly technologized society, we might have created something like the Goldberg machine without even realizing it. The sparsity method is therefore an essential technique to avoid such a situation.

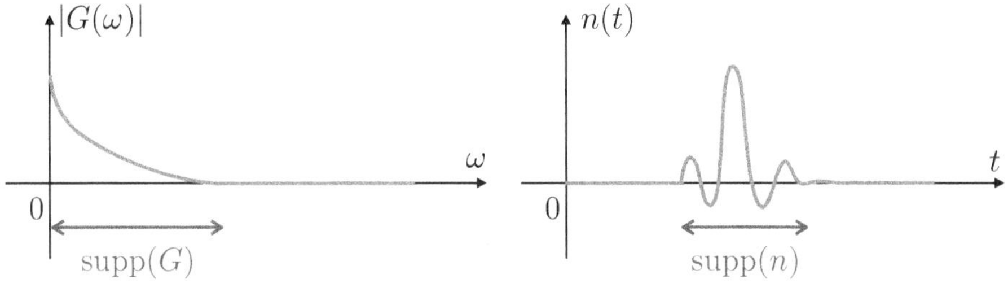

Figure 1.2: Reconstruction from noisy signal $f(t) = g(t) + n(t)$: the Fourier transform $G(\omega)$ of g is band-limited in the frequency domain, and the noise n is localized in the time domain.

1.2 Optimization with ℓ^1 Norm

As we will see in this book, ℓ^1-norm optimization is one of the most important techniques for compressed sensing as an approximation of ℓ^0-norm optimization.

1.2.1 Signal reconstruction

The first study of optimization with the ℓ^1 norm for sparse solutions is found in the dissertation by Franklin Logan in 1965 [86]. Logan considered the problem of *signal reconstruction* from noisy data. In his dissertation, he showed that ℓ^1-norm minimization completely eliminates the noise when the original signal is band-limited to a certain frequency and the noise is well localized (i.e., sparse) on the time axis. More precisely, if we have a noisy observation

$$f(t) = g(t) + n(t), \quad t = 0, 1, 2, \ldots \tag{1.1}$$

where the Fourier transform $G(\omega)$ of g has its support on a low-frequency range, and the support of $n(t)$ is sufficiently short, then the ℓ^1 optimization leads to perfect reconstruction of g from f. This is called *Logan's phenomenon*. Figure 1.2 illustrates the signal assumptions (band limitation and sparsity) in Logan's phenomenon. The sparsity method by the ℓ^1-norm minimization was then extended in [37] to signal recovery when the original signal is sparse in the frequency domain.

1.2.2 Geophysics

In the field of *geophysics*, sparsity methods by the ℓ^1 optimization have been proposed since the 1970s. The structure of the strata can be estimated by generating artificial earthquakes near the ground surface and observing

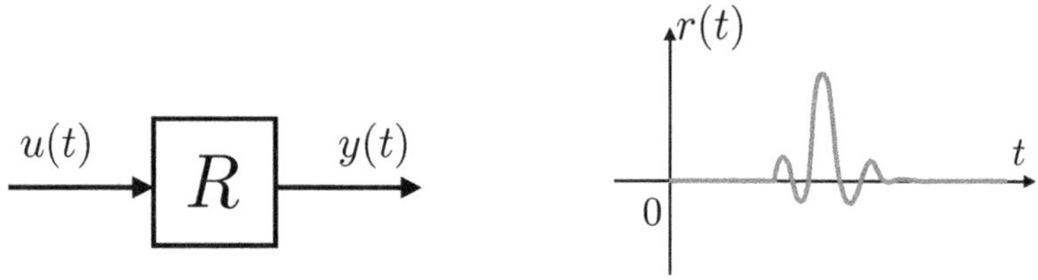

Figure 1.3: Linear system R with input $u(t)$ and output $y(t)$ (left), and its sparse impulse response $r(t)$ (right).

the reflected waves. This is a method called the *reflection seismic survey*. This is a problem of *system identification* or an *inverse problem*, where the mathematical model of the system is learned from its inputs and outputs. As shown in the left-hand diagram in Figure 1.3, we consider a linear system R with the input (the wave by the artificial earthquake) $u(t)$ and the output (the reflected wave) $y(t)$.

The problem is to find the impulse response $r(t)$ of the system R from the input/output data of $u(t)$ and $y(t)$. In the case of seismic reflection waves, the impulse response $r(t)$ can be assumed to be localized in time (see the right-hand figure in Fig. 1.3). That is, the impulse response is *sparse*. From this, the ℓ^1 regularization was proposed to reconstruct the sparse impulse response [27], [134], [146]. These are other early studies that used the idea of sparsity.

1.2.3 Neural networks

In the field of *neural networks*, the idea of sparsity has also been investigated. Since the 1980s, Masumi Ishikawa has been proposing a method to learn the structure by introducing the ℓ^1 norm regularization into the training of multilayer perceptrons [69]. He proposed sparsifying the coupling weights of the network, which can also avoid overfitting. This is a method of machine learning that takes advantage of the human brain's ability to forget, which is called *structure learning with forgetting*. The method can lead to an explainable structure of multilayer neural networks, which allows people to understand the learning results. Recently, more sophisticated methods have been proposed to obtain sparse neural networks [87], [137], [151]. In addition, the technique of *dropout* in recent deep neural networks is based on a similar idea of sparsity [49], [143].

1.2.4 Statistics

In statistics, the method called *LASSO* (Least Absolute Shrinkage and Selection Operator) is the most famous method with sparsity. Let us consider polynomial curve fitting from given data. If we can set many of the coefficients (parameters) to zero, the terms with zero coefficients will not affect the estimation at all, and we can avoid overfitting. Such a method is called a *shrinkage method* in statistics. LASSO, the shrinkage method with ℓ^1 norm regularization, was proposed by Robert Tibshirani in 1996 [147]. The idea of LASSO has been extended to *elastic net regularization* [158] with the sum of the ℓ^1 norm and the squared ℓ^2 norm as a regularized term, and *group LASSO* with the sum of weighted ℓ^2 norms for grouped vectors [155]. We will study LASSO in Chapter 3.

1.2.5 Signal processing

The first research area where sparsity methods became a hot topic is signal processing. A method called *basis pursuit* with ℓ^1 norm optimization to recover sparse signals was proposed in 1994 by Chen and Donoho [25] at the 28th Asilomar Conference on Signals, Systems, and Computers[1]. In addition, the *total variation denoising*, by using the ℓ^1 norm of the difference of a signal was proposed in 1992 by Rudin et al. [131]. More recently, Donoho et al. proposed a new theory of sensing and recovery called *compressed sensing* [36] in 2006, which is a theoretically refined version of the basis pursuit. In the same year, Candes and Tao also published a paper on this topic [19]. 2006 is the year that the current development of compressed sensing began. Compressed sensing was a topic in the fields of signal processing and information theory at that time. However, the topic is now widely attracting a lot of attention in various research fields, including systems and control.

1.3 Sparsity Methods for Systems and Control

Here we describe a brief history of sparsity methods for systems and control to provide a motivation for studying the new research topic.

[1]Later, this work was published as a journal article [24] with Saunders as a co-author.

1.3.1 Minimum-fuel control and L^1 optimization

In the field of automatic control, sparsity has been recognized for a long time. An example is the *minimum-fuel control*, which is an optimal control that minimizes the L^1 norm of control among feasible controls. The minimum-fuel control has been actively discussed in the field of control theory since the early 1960s [3]. At that time, the space race between the United States and the Soviet Union was at its peak. Minimum-fuel control has its roots in the exploration of ways to minimize rocket fuel consumption, particularly during missions from Earth to the Moon. As we will see in Chapter 8, the minimum-fuel control is a *bang-off-bang control* that takes ternary values of $\pm u_{\max}$ (the maximum amplitude that the control can produce) and *zero*, under some assumptions. When the control input is zero, the rocket undergoes inertial flight, reducing fuel consumption during this period. This is why it is called minimum-fuel control.

1.3.2 Maximum hands-off control

The L^1-optimal minimum-fuel control is shown to be equivalent to the L^0-optimal control (the sparsest control) in [104], [105] under the assumption of non-singularity. The sparse control with the minimum L^0 norm is called the *maximum hands-off control*. The mathematical properties of the maximum hands-off control have been investigated in [23], [64], [80]. This has also been extended to time-space sparse control [67], [68], time-optimal control [66], distributed control [60], [62], continuous control [110], and infinite-dimensional systems [59]. See survey papers [107], [108], [114] for detailed discussions. The maximum hands-off control will be discussed in Chapters 9–11.

1.3.3 Discrete-valued control

The bang-off-bang property of minimum-fuel control is also referred to as *discreteness*. It has been known since the 1960s that certain types of optimal control show such discreteness of control. In fact, the classical textbook [3] states that the discrete-valued control can be implemented as a few switches in the rocket cockpit (see Fig. 1.4). Clearly, such simple manual control would be impractical and dangerous for spaceflight. An automatic control system, not controlled by a human, with feedback is essential. However, the discrete-valued control expressed only by switching on and off, is very important in recent resource-aware networked control

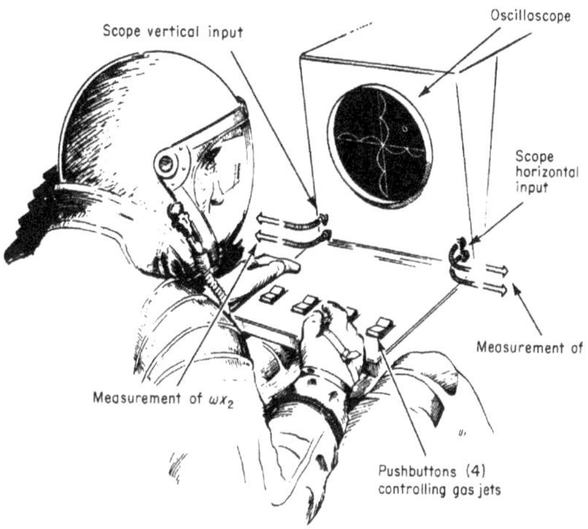

Fig. 7-62 Time-optimal manual control.

Figure 1.4: The rocket cockpit illustrated in 1960's textbook [3]. The pilot just operates the switches with the observation of the position and velocity of the rocket. This figure is from [3, p. 608, Fig. 7-62].

systems, such as the Internet of Things (IoT) or Cyber-Physical Systems (CPS). Discrete-valued control with the idea of sparsity was proposed in [61], [65]. In these papers, it is shown that the minimization of the *sum of absolute values* (SOAV) of the control enhances the discreteness. We will study the SOAV control in Section 11.2 of Chapter 11.

1.3.4 Robust control and rank minimization

The optimal control mentioned above requires a complete mathematical model of the controlled object (e.g., a rocket). However, there should be *uncertainties* in the model and parameters in reality, and how to deal with them has been a major challenge in automatic control theory. *Robust control*, a theory of control systems design that takes uncertainty into account, was actively studied in the 1980s, with H^∞ *control theory* being one of the most successful examples [156]. Some basic problems in H^∞ control boil down to the problem of finding a matrix satisfying *linear matrix inequalities* (LMIs) [12], [39]. An LMI is a convex constraint, which can be easily solved using convex optimization techniques, such as the interior point method. However, if you want to control a large-scale and high-dimensional system with a simple and much lower-dimensional controller, or if you need to treat structured uncertainties, the problem becomes LMIs with a matrix rank constraint, or rank minimization, which is much more difficult to solve since the rank constraint is non-convex[2].

The rank minimization problem is in general described as

$$\underset{X \in \mathbb{R}^{n \times n}}{\text{minimize}} \ \text{rank}(X) \ \text{subject to} \ M(X) + Q \succeq 0, \tag{1.2}$$

where $M(X)$ is a linear function of X, Q is a matrix, and the inequality $A \succeq B$ means $A - B$ is positive semidefinite. It is easily shown that the matrix rank is the number of non-zero singular values (i.e., the ℓ^0 norm), and hence this is a problem related to sparsity. As discussed in Chapter 3, the ℓ^0 norm is often approximated by the ℓ^1 norm. Namely, we instead minimize the sum of absolute values (i.e., the ℓ^1 norm) of the singular values. This is called the *nuclear norm* and denoted by $\|X\|_*$. That is, the rank minimization problem in (1.2) is approximated to the *nuclear norm minimization*:

$$\underset{X \in \mathbb{R}^{n \times n}}{\text{minimize}} \ \|X\|_* \ \text{subject to} \ M(X) + Q \succeq 0. \tag{1.3}$$

The pioneering work by Mesbahi and Papavassilopoulos [93] showed the equivalence between (1.2) and (1.3). Interestingly, this was published in 1997 prior to the theory of compressed sensing in the 2000s. For the equivalence, they used the property of *Z matrix*, which has not been considered in standard compressed sensing theory. Since then, a lot of related research has been conducted [2], [35], [44], [112], [129], [130]. In this book, we do not deal with rank minimization. Readers who are interested in rank minimization may refer to [91].

1.3.5 Resource-aware control for networked control systems

Sparsity methods have also been applied to *networked control systems*. A networked control system is a feedback control system where the communication between the controlled object and the controller is limited. Figure 1.5 shows an example of a networked control system. In this system, sensor data from the drone is sent to the computer (CPU) via a wireless communication network. Based on the information, CPU updates the control values for the attitude, speed, and acceleration of the drone, and returns the control commands to the drone via the network. A *Multi-agent system* is an important example of networked systems, where multiple autonomous

[2]The rank constraint can be equivalently transformed into a *bilinear matrix inequality* (BMI), which is also non-convex.

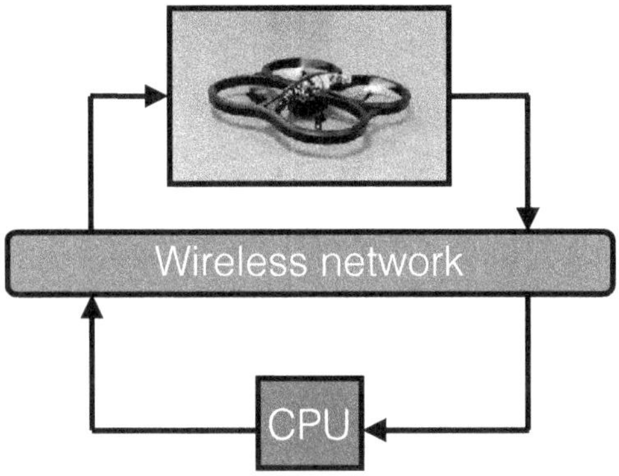

Figure 1.5: Networked Control System

agents, such as drones, collaboratively work to achieve specific goals or tasks [109].

For networked control systems, sparsity methods play an important role to realize *resource-aware control* that can significantly reduce the communication and computational burden. In [46], [78], [101], [103], [121], sparse control is proposed by using ℓ^1 norm minimization for discrete-time systems, by which we can reduce the size of control packets that are sent through rate-limited communication networks. These are finite-horizon control and to obtain feedback control, we can adapt the *receding horizon control* or the *model predictive control* formulations [32], [106], [113]. See Section 7.4 in Chapter 7 for details.

Minimum actuator placement is also an important sparsity method for resource-aware control. This is to minimize the number of actuators (or control inputs) that achieve a control objective (e.g. controllability). The problem has been discussed in [58], [70], [102], [120], [123], [125], [148].

For state feedback control, the control gain matrix is also sparsified [34], [83], [84], [96]. The obtained feedback controller is sparsely structured and the design should achieve an optimal tradeoff between closed-loop performance and sparsity. See a review paper [74] by Jovanović and Dhingra for detailed discussion on this topic. Also, we will discuss this in Section 7.3 in Chapter 7.

Part I

Compressed Sensing in Finite-dimensional Spaces

Chapter 2

What is Sparsity?

In this chapter, we explain the notion of *sparsity*, and introduce sparse representation of vectors and functions. The concepts presented here are fundamental to the remainder of this book, and should not be overlooked.

Key ideas of Chapter 2

- Sparsity of a vector is measured by its ℓ^0 norm.
- In sparse representation, a redundant dictionary of vectors is used.
- In sparse representation, the smallest number of vectors are automatically chosen from a redundant dictionary that represent a given vector, which is achieved by ℓ^0 optimization.
- The exhaustive search to solve the ℓ^0 optimization requires computational time that exponentially increases as the problem size increases.

2.1 Redundant Dictionary

Let us consider the three-dimensional vector space $\mathbb{R}^3$. The *standard basis* for $\mathbb{R}^3$ is formed with the following three unit vectors:

$$e_1 = \begin{bmatrix} 1 \\ 0 \\ 0 \end{bmatrix}, \quad e_2 = \begin{bmatrix} 0 \\ 1 \\ 0 \end{bmatrix}, \quad e_3 = \begin{bmatrix} 0 \\ 0 \\ 1 \end{bmatrix}. \tag{2.1}$$

By using this basis, any three-dimensional vector $\boldsymbol{y} \in \mathbb{R}^3$ can be represented as

$$\boldsymbol{y} = \begin{bmatrix} y_1 \\ y_2 \\ y_3 \end{bmatrix} = y_1\boldsymbol{e}_1 + y_2\boldsymbol{e}_2 + y_3\boldsymbol{e}_3. \tag{2.2}$$

In general, if you choose three linearly independent vectors $\boldsymbol{\phi}_1$, $\boldsymbol{\phi}_2$, and $\boldsymbol{\phi}_3$ from $\mathbb{R}^3$, then they form a basis for $\mathbb{R}^3$. That is, for any vector $\boldsymbol{y} \in \mathbb{R}^3$, there exist unique real numbers β_1, β_2, and β_3 such that

$$\boldsymbol{y} = \beta_1\boldsymbol{\phi}_1 + \beta_2\boldsymbol{\phi}_2 + \beta_3\boldsymbol{\phi}_3 \tag{2.3}$$

holds. Moreover, if $\boldsymbol{\phi}_1$, $\boldsymbol{\phi}_2$, and $\boldsymbol{\phi}_3$ are unit vectors and orthogonal to each other, that is, if they satisfy

$$\langle \boldsymbol{\phi}_i, \boldsymbol{\phi}_j \rangle = \boldsymbol{\phi}_i^\top \boldsymbol{\phi}_j = \begin{cases} 1, & i = j \\ 0, & i \neq j \end{cases}, \quad i, j = 1, 2, 3, \tag{2.4}$$

where $\langle \cdot, \cdot \rangle$ is the ℓ^2 inner product (see also Section 2.3), then $\boldsymbol{\phi}_1$, $\boldsymbol{\phi}_2$, and $\boldsymbol{\phi}_3$ form an *orthonormal basis* for $\mathbb{R}^3$, and the coefficients β_1, β_2, and β_3 can be obtained by the inner product

$$\beta_i = \langle \boldsymbol{\phi}_i, \boldsymbol{y} \rangle = \boldsymbol{\phi}_i^\top \boldsymbol{y}, \quad i = 1, 2, 3. \tag{2.5}$$

Exercise 2.1. How do you obtain the coefficients β_1, β_2, and β_3 in (2.3), when $\boldsymbol{\phi}_1$, $\boldsymbol{\phi}_2$, and $\boldsymbol{\phi}_3$ are linearly independent but they do not form an orthonormal basis?

Let us consider another basis for $\mathbb{R}^3$ with the following three linearly independent vectors:

$$\boldsymbol{\phi}_1 = \boldsymbol{e}_1 + \boldsymbol{e}_2 = \begin{bmatrix} 1 \\ 1 \\ 0 \end{bmatrix}, \quad \boldsymbol{\phi}_2 = \boldsymbol{e}_2 + \boldsymbol{e}_3 = \begin{bmatrix} 0 \\ 1 \\ 1 \end{bmatrix}, \quad \boldsymbol{\phi}_3 = \boldsymbol{e}_3 + \boldsymbol{e}_1 = \begin{bmatrix} 1 \\ 0 \\ 1 \end{bmatrix}. \tag{2.6}$$

Combining these with the unit vectors in (2.1), let us form a set of 6 vectors $\{\boldsymbol{e}_1, \boldsymbol{e}_2, \boldsymbol{e}_3, \boldsymbol{\phi}_1, \boldsymbol{\phi}_2, \boldsymbol{\phi}_3\}$. Figure 2.1 shows these 6 vectors. With these vectors, consider the following representation of vector $\boldsymbol{y} \in \mathbb{R}^3$:

$$\boldsymbol{y} = \sum_{i=1}^{3} \alpha_i \boldsymbol{e}_i + \sum_{i=1}^{3} \beta_i \boldsymbol{\phi}_i. \tag{2.7}$$

This is a *redundant* representation, and there are infinitely many solutions for α_i and β_i ($i = 1, 2, 3$) to satisfy (2.7). For example, for $\boldsymbol{y} = [y_1, y_2, y_3]^\top$, we have two solutions

$$(\alpha_1, \alpha_2, \alpha_3, \beta_1, \beta_2, \beta_3) = (y_1, y_2, y_3, 0, 0, 0), \tag{2.8}$$

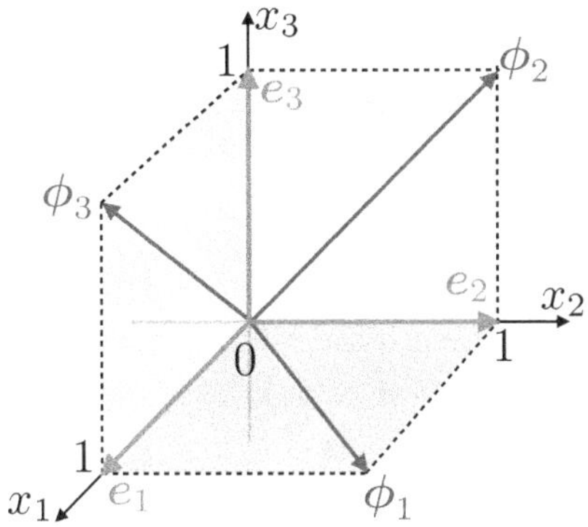

Figure 2.1: 6 vectors $e_1, e_2, e_3, \phi_1, \phi_2, \phi_3$ in $\mathbb{R}^3$

and

$$(\alpha_1, \alpha_2, \alpha_3, \beta_1, \beta_2, \beta_3) = (-y_3, -y_1, -y_2, y_1, y_2, y_3). \tag{2.9}$$

Now, let us consider a situation where the cost to keep the values of the non-zero coefficients is very expensive due to an expensive memory device for example. Then we want to minimize the number of non-zero coefficients to reduce the cost. Let us consider a vector $\boldsymbol{y} \in \mathbb{R}^3$ on the plane spanned by $\boldsymbol{e}_1$ and $\boldsymbol{\phi}_2$. For this vector, we have the following solution:

$$\boldsymbol{y} = \alpha_1 \boldsymbol{e}_1 + \beta_2 \boldsymbol{\phi}_2, \tag{2.10}$$

This expression has a smaller number of non-zero coefficients than (2.8). This is a trivial example and the cost of (2.10) is almost the same as that of (2.8). However, if we can find just 10^2 non-zero coefficients for a 10^6 (one million) dimensional vector, the cost will be dramatically reduced. Such a technology is often called *data compression*, which is one of the biggest motivations of *sparse representation*.

Example 2.1. The four cardinal directions form a redundant system to represent a direction in $\mathbb{R}^2$. We say, for example, "Go southwest" not "Go minus-north-minus-east" although the two are mathematically equivalent. □

Example 2.2. Imagine that you want to tell a foreigner about an elephant. The foreigner cannot speak English but has a small dictionary with 3,000 words, which does not include the word 'elephant.' You might say, 'There is an animal that is the largest living land animal and has a long nose. Many of them live in the African savanna.' Then the foreigner might ask, 'What is a savanna?' since the word is not in their dictionary. However, if the foreigner has a larger dictionary that contains more than 1 million words, you could simply say, 'There is an elephant.' Some English teachers claim you only need to memorize these 3,000 words for conversation, but actually, 3,000 words are not enough at all for simple expression. □

Let us formulate this problem of sparse representation in a general form. Let us consider m-dimensional vector space $\mathbb{R}^m$, and a set of vectors $\{\phi_1, \phi_2, \ldots, \phi_n\}$ in $\mathbb{R}^m$, where $n > m$. For a given vector $\boldsymbol{y} \in \mathbb{R}^m$, we find coefficients $\alpha_1, \alpha_2, \ldots, \alpha_n$ such that

$$\boldsymbol{y} = \sum_{i=1}^{n} \alpha_i \boldsymbol{\phi}_i. \tag{2.11}$$

We assume that m vectors in $\{\phi_1, \phi_2, \ldots, \phi_n\}$ are linearly independent. We call such a set of vectors $\{\phi_1, \phi_2, \ldots, \phi_n\}$ a *dictionary* (recall Example 2.2), and the elements $\boldsymbol{\phi}_1, \boldsymbol{\phi}_2, \ldots, \boldsymbol{\phi}_n$ *atoms*[1]. Note that the size n of the dictionary is larger than the size m of vector $\boldsymbol{y}$. We call such a dictionary a *redundant dictionary*, or *over-complete dictionary*.

Define a matrix Φ and a vector $\boldsymbol{x}$ as

$$\Phi \triangleq \begin{bmatrix} \phi_1 & \phi_2 & \ldots & \phi_n \end{bmatrix} \in \mathbb{R}^{m \times n}, \quad \boldsymbol{x} \triangleq \begin{bmatrix} \alpha_1 \\ \alpha_2 \\ \vdots \\ \alpha_n \end{bmatrix} \in \mathbb{R}^n. \tag{2.12}$$

Then, the equation (2.11) can be equivalently written as

$$\Phi \boldsymbol{x} = \boldsymbol{y}. \tag{2.13}$$

The matrix Φ is called a *dictionary matrix*, or a *measurement matrix*. Since the dictionary is redundant, the matrix Φ is a *fat* matrix, that is, the number of columns is larger than the number of rows. Our problem is now described as follows:

Problem 2.1 (Sparse Representation). Given a vector $\boldsymbol{y} \in \mathbb{R}^m$ and a dictionary $\{\phi_1, \phi_2, \ldots, \phi_n\}$. Find the simplest representation of $\boldsymbol{y}$ that satisfies (2.13).

In the next section, we discuss this problem with a fat matrix.

[1]We do not call them *words*.

2.2 Underdetermined Systems

Let us consider the following system of linear equations with unknowns x_1, x_2, and x_3:

$$\begin{aligned} x_1 + x_2 + x_3 &= 3 \\ x_1 - x_3 &= 0 \end{aligned} \tag{2.14}$$

Now there are three unknowns and two equations, and it is easily seen that there are infinitely many solutions. To represent all solutions, we use parametrization. All solutions to (2.14) are parametrized as

$$x_1 = t, \quad x_2 = -2t + 3, \quad x_3 = t, \tag{2.15}$$

where $t \in \mathbb{R}$ is a parameter. We call such a system of equations an *underdetermined system*, where the number of unknowns is larger than the number of equations.

An underdetermined system is something like insufficient proofs for a detective to determine one among many suspects. For a detective, say Conan Edogawa[2], the two proofs (equations) in (2.14) are insufficient and he should seek one more proof to reveal the unique solution to the case. Thanks to his investigation, a proof was found, which said *"the criminal is the smallest one among the suspects."* This is actually a conclusive proof by which one can choose just one suspect. Let us find the smallest solution among the candidates in (2.15). We use the ℓ^2 norm as a measure of the size, and we find the smallest ℓ^2-norm solution as follows. First, from (2.15), we have

$$\begin{aligned} \|\boldsymbol{x}\|_2^2 &= x_1^2 + x_2^2 + x_3^2 \\ &= t^2 + (-2t + 3)^2 + t^2 \\ &= 6(t-1)^2 + 3. \end{aligned} \tag{2.16}$$

Then we can choose $t = 1$, and from (2.15), the solution is uniquely chosen as $(x_1, x_2, x_3) = (1, 1, 1)$. *Case closed.*

Let us generalize the above discussion. We consider a system of linear equations in a matrix form as

$$\Phi \boldsymbol{x} = \boldsymbol{y}. \tag{2.17}$$

For example, the system in (2.14) can be represented in the matrix form

[2]See: `https://en.wikipedia.org/wiki/Case_Closed`

(2.17) with

$$\Phi = \begin{bmatrix} 1 & 1 & 1 \\ 1 & 0 & -1 \end{bmatrix}, \quad \boldsymbol{x} = \begin{bmatrix} x_1 \\ x_2 \\ x_3 \end{bmatrix}, \quad \boldsymbol{y} = \begin{bmatrix} 3 \\ 0 \end{bmatrix}. \tag{2.18}$$

We assume the size of matrix Φ is $m \times n$ where $m < n$, that is, we consider an underdetermined system of equations. We also assume that there are m column vectors in $\{\boldsymbol{\phi}_1, \ldots, \boldsymbol{\phi}_n\}$ that are linearly independent. In other words, we assume Φ has *full row rank*. Note that a matrix $\Phi \in \mathbb{R}^{m \times n}$ is said to have full row rank if Φ is *surjective*, or

$$\operatorname{rank}(\Phi) = m. \tag{2.19}$$

If $\operatorname{rank}(\Phi) < m$, then there exist redundant linear equations (i.e., there is at least one equation that is a linear combination of other equations). For example, the following system of equations

$$\begin{aligned} x_1 + x_2 + x_3 &= 3 \\ x_1 - x_3 &= 0 \\ x_1 - x_3 &= 0 \end{aligned} \tag{2.20}$$

is redundant and the rank is $2 < 3$. We here assume such redundancy should be eliminated beforehand.

If Φ has full row rank, or $\operatorname{rank}(\Phi) = m$, then for any vector $\boldsymbol{y} \in \mathbb{R}^m$, there exists at least one solution $\boldsymbol{x}$ that satisfies the linear equation (2.17). Now, we seek *all* the solutions to (2.17). For this, we define the *kernel* (or *null space*) of matrix Φ by

$$\ker(\Phi) \triangleq \{\boldsymbol{x} \in \mathbb{R}^n : \Phi\boldsymbol{x} = \boldsymbol{0}\}. \tag{2.21}$$

Note that $\ker(\Phi)$ is a linear subspace in $\mathbb{R}^n$, that is, if $\boldsymbol{x}_1, \boldsymbol{x}_2 \in \ker(\Phi)$, then $a_1\boldsymbol{x}_1 + a_2\boldsymbol{x}_2 \in \ker(\Phi)$ holds for any $a_1, a_2, \in \mathbb{R}$. Then, we introduce the *dimension theorem* in linear algebra.

Theorem 2.1 (dimension theorem). For any matrix $\Phi \in \mathbb{R}^{m \times n}$,

$$\operatorname{rank}(\Phi) + \dim \ker(\Phi) = n \tag{2.22}$$

holds.

From the dimension theorem, the dimension of $\ker(\Phi)$ is $n - m$. Since $n > m$, the kernel, which is a linear subspace in $\mathbb{R}^n$, has at least one dimension. That is, there exist infinitely many vectors in $\ker(\Phi)$.

Let $\boldsymbol{x}_0 \in \mathbb{R}^n$ be a *particular solution* to (2.17). Then, all the solutions to the linear equation (2.17) can be represented by the sum of the particular solution $\boldsymbol{x}_0$ and a *free parameter* $\boldsymbol{z} \in \ker(\Phi)$, that is,

$$\boldsymbol{x} = \boldsymbol{x}_0 + \boldsymbol{z}, \quad \boldsymbol{z} \in \ker(\Phi). \tag{2.23}$$

From this, it follows that there exist infinitely many solutions to (2.17).

Exercise 2.2. Show that the vector $\boldsymbol{x}$ in (2.23) is the solution to the equation (2.17).

The problem of sparse representation (Problem 2.1) is to find a solution $\boldsymbol{x}$ to (2.17) that has the simplest representation or the smallest number of non-zero elements. Let us consider this problem more precisely in the next section.

2.3 The ℓ^0 Norm

We here review the notion of a norm in a finite-dimensional vector space, and then introduce the ℓ^0 norm that defines the sparsity of a vector.

First, let us recall the definition of a norm in $\mathbb{R}^n$.

Definition 2.1. A *norm* $\|\boldsymbol{x}\| : \mathbb{R}^n \to [0, \infty)$ is a nonnegative function that satisfies the following properties:

1. For any vector $\boldsymbol{x} \in \mathbb{R}^n$ and any number $\alpha \in \mathbb{R}$, $\|\alpha\boldsymbol{x}\| = |\alpha|\|\boldsymbol{x}\|$.
2. For any $\boldsymbol{x}, \boldsymbol{y} \in \mathbb{R}^n$, $\|\boldsymbol{x} + \boldsymbol{y}\| \leq \|\boldsymbol{x}\| + \|\boldsymbol{y}\|$.
3. $\|\boldsymbol{x}\| = 0 \iff \boldsymbol{x} = \boldsymbol{0}$.

A well-known norm in $\mathbb{R}^n$ is the ℓ^2 norm (or the *Euclidean norm*). For a vector $\boldsymbol{x} = [x_1, x_2, \ldots, x_n]^\top \in \mathbb{R}^n$, the ℓ^2 norm is defined by

$$\|\boldsymbol{x}\|_2 \triangleq \sqrt{x_1^2 + x_2^2 + \cdots + x_n^2}. \tag{2.24}$$

The ℓ^2 norm is also given by

$$\|\boldsymbol{x}\|_2 = \sqrt{\langle \boldsymbol{x}, \boldsymbol{x} \rangle}, \tag{2.25}$$

where $\langle \cdot, \cdot \rangle$ is the ℓ^2 *inner product* (or *Euclidean inner product*) in $\mathbb{R}^n$ defined by

$$\langle \boldsymbol{x}, \boldsymbol{y} \rangle \triangleq \boldsymbol{x}^\top \boldsymbol{y} = \sum_{i=1}^{n} x_i y_i. \tag{2.26}$$

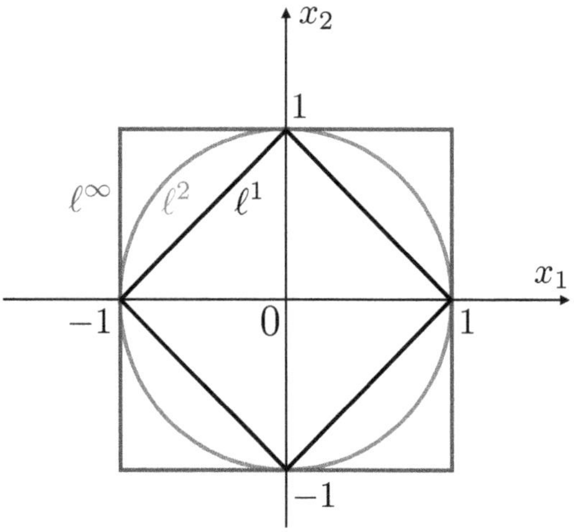

Figure 2.2: Contour curves ($\|\boldsymbol{x}\|_p = 1$) of ℓ^1, ℓ^2, ℓ^∞ norms.

Exercise 2.3. Confirm the ℓ^2 norm $\|\boldsymbol{x}\|_2$ defined in (2.24) satisfies the three properties in Definition 2.1.

Beyond the ℓ^2 norm, an infinite variety of norms can be defined for $\mathbb{R}^n$. A generalization of the ℓ^2 norm in (2.24) is the ℓ^p *norm* with $p \in [1, \infty)$, defined by

$$\|\boldsymbol{x}\|_p \triangleq \left(\sum_{i=1}^{n} |x_i|^p \right)^{1/p}. \tag{2.27}$$

The most important norm in this book is the ℓ^1 *norm* with $p = 1$ in (2.27). The ℓ^1 norm is described as the sum of the absolute values of the elements in a vector, that is,

$$\|\boldsymbol{x}\|_1 = \sum_{i=1}^{n} |x_i|. \tag{2.28}$$

The limit of (2.27) as $p \to \infty$ is called the ℓ^∞ *norm* (or the *maximum norm*), defined by

$$\|\boldsymbol{x}\|_\infty \triangleq \max_{i=1,2,\dots,n} |x_i|. \tag{2.29}$$

Exercise 2.4. Prove that for any $\boldsymbol{x} \in \mathbb{R}^n$,

$$\|\boldsymbol{x}\|_\infty = \lim_{p \to \infty} \|\boldsymbol{x}\|_p. \tag{2.30}$$

Figure 2.2 shows the contour curves that satisfy $\|\boldsymbol{x}\|_p = 1$ for $p = 1, 2$, and ∞ in $\mathbb{R}^2$. The contour of the ℓ^2 norm is a unit circle centered at the origin. The contour of the ℓ^∞ norm is a unit square centered at the origin, and touches the ℓ^2 circle at $(1,0)$, $(0,1)$, $(-1,0)$, and $(0,-1)$. The shape of the contour of the ℓ^1 norm is very important for sparse representation. This diamond-shaped contour has four corners on the x_1 and x_2 axes. This property gives an intuitive explanation of the relation between ℓ^1 norm and sparsity (see Section 3.2).

Now, let us define the ℓ^0 norm. Consider a vector $\boldsymbol{x} = [x_1, x_2, \ldots, x_n]^\top \in \mathbb{R}^n$. Define the *support* of $\boldsymbol{x}$ by

$$\operatorname{supp}(\boldsymbol{x}) \triangleq \{i \in \{1, 2, \ldots, n\} : x_i \neq 0\}. \tag{2.31}$$

The support of $\boldsymbol{x}$ is the set of indices on which the elements of $\boldsymbol{x}$ are nonzero. By using the support, the ℓ^0 *norm* is defined by

$$\|\boldsymbol{x}\|_0 \triangleq \#(\operatorname{supp}(\boldsymbol{x})), \tag{2.32}$$

where $\#(\operatorname{supp}(\boldsymbol{x}))$ is the number of elements in the finite set $\operatorname{supp}(\boldsymbol{x})$. Namely, the ℓ^0 norm counts the number of nonzero elements in $\boldsymbol{x}$.

It is notable that the ℓ^0 norm does not satisfy the first property in Definition 2.1. For example, a nonzero vector $\boldsymbol{x} \in \mathbb{R}^n$ has the same ℓ^0 norm as $2\boldsymbol{x}$. This implies that

$$\|2\boldsymbol{x}\|_0 = \|\boldsymbol{x}\|_0 \neq 2\|\boldsymbol{x}\|_0, \tag{2.33}$$

whenever $\boldsymbol{x} \neq 0$. Strictly speaking, the ℓ^0 norm is not a norm, and hence we sometimes call it as ℓ^0 *pseudo-norm* or *cardinality*. However, we use the term "ℓ^0 norm" as often used in the literature. Note that by definition, the second and third properties in Definition 2.1 hold, that is,

$$\|\boldsymbol{x} + \boldsymbol{y}\|_0 \leq \|\boldsymbol{x}\|_0 + \|\boldsymbol{y}\|_0 \tag{2.34}$$

and

$$\|\boldsymbol{x}\|_0 = 0 \iff \boldsymbol{x} = \boldsymbol{0}. \tag{2.35}$$

Finally, we define the sparsity of a vector by using the ℓ^0 norm. A vector $\boldsymbol{x} \in \mathbb{R}^n$ is said to be *sparse* if the ℓ^0 norm $\|\boldsymbol{x}\|_0$ is sufficiently small compared to the dimension n. The notion of sparsity is important in this book.

Exercise 2.5. Prove that for any $\boldsymbol{x}, \boldsymbol{y} \in \mathbb{R}^n$,

$$\|\boldsymbol{x} + \boldsymbol{y}\|_0 \leq \|\boldsymbol{x}\|_0 + \|\boldsymbol{y}\|_0 \tag{2.36}$$

holds.

Exercise 2.6. Let $\boldsymbol{x}, \boldsymbol{y} \in \mathbb{R}^n$. When does the following equality hold?

$$\|\boldsymbol{x} + \boldsymbol{y}\|_0 = \|\boldsymbol{x}\|_0 + \|\boldsymbol{y}\|_0. \tag{2.37}$$

The problem of sparse representation (Problem 2.1) is finding the sparsest solution among infinitely many solutions to the linear equation in (2.17). This problem is mathematically formulated by using the ℓ^0 norm introduced above. That is, we seek the smallest ℓ^0-norm solution to (2.17). This is formulated as a mathematical optimization problem as follows:

Problem 2.2 (Sparse representation). Given a vector $\boldsymbol{y} \in \mathbb{R}^m$ and a full-row-rank matrix $\Phi \in \mathbb{R}^{m \times n}$. Find the optimizer $\boldsymbol{x}^*$ of the optimization problem:

$$\underset{\boldsymbol{x} \in \mathbb{R}^n}{\text{minimize}} \ \|\boldsymbol{x}\|_0 \ \text{ subject to } \ \Phi \boldsymbol{x} = \boldsymbol{y}. \tag{2.38}$$

We call this the ℓ^0 *optimization*.

2.4 Group Testing

In this section, we consider a real example of compressed sensing called *group testing*, which is one of the first attempts to apply a sparsity method to a scientific problem. Group testing was proposed by Robert Dorfman in 1948 as a problem of finding an infected individual among a large number of patients in a small number of blood tests [38].

For example, suppose that only one of eight patients is infected with a disease, which can be detected by examining the blood. Now, we have eight blood samples from the eight patients. Since blood testing is expensive and time-consuming, we want to identify the infected individual as few times as possible. In this case, there is a good way to do this (see also Figure 2.3).

- (TEST 1) We first divide the blood of the eight patients into two groups of four patients, and take a little bit of blood from each of the eight patients, and mix it for each group. Since there is only one infected individual, the blood from either group will test positive.
- (TEST 2) Divide the group that tested positive into two groups of two patients, and do the same thing as above. At this point, the number of suspicious individual has been narrowed down to two.
- (TEST 3) Finally, by examining the blood of the two individuals separately, the infected individual can be uniquely identified.

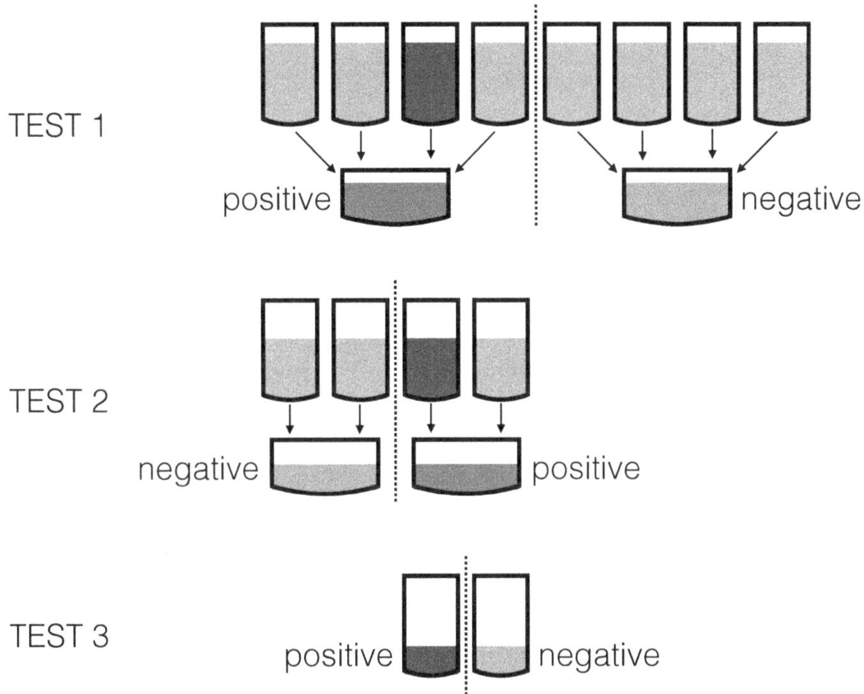

Figure 2.3: Group testing from eight blood samples

By this method, it is possible to identify an infected individual in three tests, whereas eight tests would be required for an individual blood test. In general, according to the above method, if there is only one infected individual among 2^T people, we can identify the infected individual in less than T tests. For example, for 1,024 patients, only 10 tests are needed to identify the infected individual. We can see that group testing can dramatically reduce the number of tests compared to testing all patients' blood individually. Then we consider a sophisticated method like this in a general situation where a few people in 100,000, for example, are infected, instead of examining the blood of 100,000 people individually. This is the problem of group testing.

Now let us describe the problem of group testing in detail. Let n be the number of people to be tested. Define a variable x_i representing whether the i-th individual ($i \in \{1, 2, \ldots, n\}$) is infected or not as

$$x_i \triangleq \begin{cases} 1, & \text{if the } i\text{-th individual is infected,} \\ 0, & \text{otherwise.} \end{cases} \tag{2.39}$$

Define an n-dimensional binary vector that takes values of 0 or 1 as

$$\boldsymbol{x} \triangleq [x_1, x_2, \ldots, x_n]^\top \in \{0,1\}^n, \tag{2.40}$$

where $\{0,1\}^n$ is the set of n-dimensional vectors whose elements are 0 or 1. The problem here is to find this n-dimensional binary vector. Of course, if we examine each one of them individually, we can determine the vector $\boldsymbol{x}$ with n tests, but here we want to identify $\boldsymbol{x}$ with a much smaller number of tests.

To formulate the process of group testing, we define the *testing vector* $\boldsymbol{a} \in \{0,1\}^n$, where one-valued elements in $\boldsymbol{a}$ indicate which blood samples are tested. For example, if $\boldsymbol{x} = [0,1,0,1,0,0,0,0]^\top$ and $\boldsymbol{a} = [1,1,1,1,0,0,0,0]^\top$, then we test the first four elements in $\boldsymbol{x}$, and the result is

$$\langle \boldsymbol{a}, \boldsymbol{x} \rangle = 2. \tag{2.41}$$

This means that there are two positives among the first four individuals. Here we assume the blood test is very precise so that the number of positives in mixed blood can be detected.

We here assume the number of tests is m, which is much less than n, the number of individuals. We also assume the testing vector $\boldsymbol{a}_j$, $j = 1, 2, \ldots, m$, are previously given. This is called the *non-adaptive group testing.* On the other hand, if, for example, $\boldsymbol{a}_2$ depends on the result $\langle \boldsymbol{a}_1, \boldsymbol{x} \rangle$ as in the example above (see Figure 2.3), then this is called the *adaptive group testing.* From this formulation, the j-th result ($j = 1, 2, \ldots, m$) of group testing is given by

$$y_j = \langle \boldsymbol{a}_j, \boldsymbol{x} \rangle, \quad j = 1, 2, \ldots, m. \tag{2.42}$$

Then, define the following matrix and vector:

$$\Phi \triangleq \begin{bmatrix} \boldsymbol{a}_1^\top \\ \boldsymbol{a}_2^\top \\ \vdots \\ \boldsymbol{a}_m^\top \end{bmatrix} \in \mathbb{R}^{m \times n}, \quad \boldsymbol{y} \triangleq \begin{bmatrix} y_1 \\ y_2 \\ \vdots \\ y_m \end{bmatrix} \in \mathbb{R}^m. \tag{2.43}$$

The problem of group testing is to find a vector $\boldsymbol{x} \in \{0,1\}^n$ that satisfies the linear equation

$$\Phi \boldsymbol{x} = \boldsymbol{y}. \tag{2.44}$$

The matrix Φ is called the *testing matrix*, or the *measurement matrix*, and the vector $\boldsymbol{y}$ is called the *measurement vector*.

If the matrix $\Phi \in \mathbb{R}^{m \times n}$ is fixed (i.e., non-adaptive group testing), and has full row rank (i.e., $\text{rank}(\Phi) = m$), then the equation $\Phi \boldsymbol{x} = \boldsymbol{y}$ is underdetermined and has infinitely many solutions since $m < n$. To obtain the solution uniquely, we further assume that there are just a few positives in n individuals. In other words, the ℓ^0 norm of the solution $\boldsymbol{x}$ is much smaller than n, that is, the solution $\boldsymbol{x}$ is sparse. Then our problem finding a sparse $\boldsymbol{x}$ is formulated as the ℓ^0 optimization (2.38).

2.5 Exhaustive Search

In this section, we show a direct method to solve the ℓ^0 optimization problem (2.38), called an *exhaustive search* (or *brute-force search*). Let $\boldsymbol{\phi}_i \in \mathbb{R}^m$ $(i = 1, 2, \ldots, n)$ denote the i-th column vector in matrix Φ, that is,

$$\Phi \triangleq \begin{bmatrix} \boldsymbol{\phi}_1 & \boldsymbol{\phi}_2 & \ldots & \boldsymbol{\phi}_n \end{bmatrix} \in \mathbb{R}^{m \times n}. \tag{2.45}$$

The following shows the procedure of the exhaustive search for (2.38).

1. If $\boldsymbol{y} = \boldsymbol{0}$, then output $\boldsymbol{x}^* = \boldsymbol{0}$ as the optimal solution and quit. Otherwise, proceed to the next step.

2. Find a vector $\boldsymbol{x}$ with $\|\boldsymbol{x}\|_0 = 1$ that satisfies the equation $\boldsymbol{y} = \Phi \boldsymbol{x}$. That is, set

$$\boldsymbol{x}_1 \triangleq \begin{bmatrix} x_1 \\ 0 \\ \vdots \\ 0 \end{bmatrix}, \quad \boldsymbol{x}_2 \triangleq \begin{bmatrix} 0 \\ x_2 \\ 0 \\ \vdots \\ 0 \end{bmatrix}, \ldots, \quad \boldsymbol{x}_n \triangleq \begin{bmatrix} 0 \\ \vdots \\ 0 \\ x_n \end{bmatrix}, \tag{2.46}$$

   and search $x_i \in \mathbb{R}$ $(i = 1, 2, \ldots, n)$ that satisfies

$$\boldsymbol{y} = \Phi \boldsymbol{x}_i = x_i \boldsymbol{\phi}_i. \tag{2.47}$$

   If a solution exists for some i, output $\boldsymbol{x}^* = \boldsymbol{x}_i$ as the solution and quit. Otherwise, proceed to the next step.

3. Find a vector $\boldsymbol{x}$ with $\|\boldsymbol{x}\|_0 = 2$ that satisfies the equation $\boldsymbol{y} = \Phi \boldsymbol{x}$.

That is, set

$$\boldsymbol{x}_{1,2} \triangleq \begin{bmatrix} x_1 \\ x_2 \\ 0 \\ \vdots \\ 0 \end{bmatrix}, \quad \boldsymbol{x}_{1,3} \triangleq \begin{bmatrix} x_1 \\ 0 \\ x_3 \\ 0 \\ \vdots \\ 0 \end{bmatrix}, \ldots, \quad \boldsymbol{x}_{n-1,n} \triangleq \begin{bmatrix} 0 \\ \vdots \\ 0 \\ x_{n-1} \\ x_n \end{bmatrix} \tag{2.48}$$

and search $x_i, x_j \in \mathbb{R}$ $(i, j = 1, 2, \ldots, n)$ that satisfies

$$\boldsymbol{y} = \Phi \boldsymbol{x}_{i,j} = x_i \boldsymbol{\phi}_i + x_j \boldsymbol{\phi}_j. \tag{2.49}$$

If a solution exists for some i, j, then output $\boldsymbol{x}^* = \boldsymbol{x}_{i,j}$ and quit. Otherwise, proceed to the next step.

4. Do similar procedures for $\|\boldsymbol{x}\|_0 = k$, $k = 3, 4, \ldots, m$ until a solution is found.

The exhaustive search can find the optimal solution $\boldsymbol{x}^*$ (if it exists) within a finite number of steps, while the worst-case scenario requires $k = m$ steps.

Next, we investigate the exhaustive search in detail. For a vector $\boldsymbol{x} = [x_1, x_2, \ldots, x_n]^\top$ and an index set $S \subset \{1, 2, \ldots, n\}$, we denote by $\boldsymbol{x}_S \in \mathbb{R}^{\#(S)}$ the restriction of $\boldsymbol{x}$ to the indices in S, where $\#(S)$ is the number of elements in S. For example, for $\boldsymbol{x} = [x_1, x_2, x_3, x_4, x_5, x_6]^\top$ and $S = \{1, 2, 5\}$, we have

$$\boldsymbol{x}_S = \begin{bmatrix} x_1 \\ x_2 \\ x_5 \end{bmatrix} \in \mathbb{R}^3. \tag{2.50}$$

More generally, for $\boldsymbol{x} = [x_1, x_2, \ldots, x_n]^\top$ and the index set

$$S = \{i_1, i_2, \ldots, i_k\}, \quad k \in \{1, 2, \ldots, n\}, \tag{2.51}$$

where $1 \leq i_1 < i_2 < \cdots < i_k \leq n$, we have

$$\boldsymbol{x}_S = \begin{bmatrix} x_{i_1} \\ x_{i_2} \\ \vdots \\ x_{i_k} \end{bmatrix} \in \mathbb{R}^k. \tag{2.52}$$

Also, for matrix $\Phi = [\phi_1, \phi_2, \ldots, \phi_n] \in \mathbb{R}^{m\times n}$ with $\phi_i \in \mathbb{R}^m$, $i = 1, 2, \ldots, n$ and the index set in (2.51), we define

$$\Phi_S = [\phi_{i_1}, \phi_{i_2}, \ldots, \phi_{i_k}] \in \mathbb{R}^{m\times k}. \tag{2.53}$$

Using this notation, we can formulate the exhaustive search algorithm for the ℓ^0 optimization (2.38) as follows: First check if $\boldsymbol{y} = \boldsymbol{0}$. In this case, the solution is $\boldsymbol{x}^* = \boldsymbol{0}$. Otherwise, take each subset S of the index set $\{1, 2, \ldots, n\}$ from $\#(S) = 1$ to $\#(S) = m$, and solve the following equation

$$\boldsymbol{y} = \Phi_S \boldsymbol{x}_S. \tag{2.54}$$

If there is a solution to (2.54), then using the solution $\boldsymbol{x}_S = [x_{i_1}, \ldots, x_{i_k}]^\top$, set $\boldsymbol{x}^* = [x_1^*, x_2^*, \ldots, x_n^*]^\top$ where

$$x_i^* = \begin{cases} x_i, & i \in S, \\ 0 & i \notin S. \end{cases} \tag{2.55}$$

This is the sparsest solution and we have $\|\boldsymbol{x}^*\|_0 = k$. We summarize the exhaustive search algorithm.

Exhaustive search algorithm for ℓ^0 optimization (2.38)

1. If $\boldsymbol{y} = \boldsymbol{0}$ then output $\boldsymbol{x}^* = \boldsymbol{0}$ and quit. Otherwise, proceed to the next step.
2. $k := 1$.
3. For each subset $S \subset \{1, 2, \ldots, n\}$ with $\#(S) = k$, do
   - Check if equation $\boldsymbol{y} = \Phi_S \boldsymbol{x}_S$ has a solution.
   - If it exists, output $\boldsymbol{x}^*$ defined in (2.55) and quit.
4. $k := k + 1$. Return to 3.

We should notice that with the exhaustive search method, the computation time to find a solution grows exponentially with problem size (i.e. m). For example, in image processing, the dimension becomes millions or larger, and the exhaustive search is not useful at all.

The above problem is also known as *combinatorial optimization*, which is in general hard to solve for large-scale problems. In the following chapters, we investigate efficient algorithms for such a hard problem of sparse optimization.

Example 2.3. Let us consider the problem of group testing discussed in Section 2.4. Suppose we conduct $m = 100$ tests on significantly more than 100 individuals. Then we solve the optimization problem (2.38) with fixed measurement matrix Φ and obtained measurement vector $\boldsymbol{y} \in \mathbb{R}^{100}$. By the exhaustive search, the number of iterations at the worst case is $2^m = 2^{100} \approx 1.3 \times 10^{30}$. Suppose that you can use a supercomputer that can do one iteration of the exhaustive search algorithm in 10^{-15} seconds. Then, it takes $1.3 \times 10^{30} \times 10^{-15} \approx 30$ million years at the worst case! □

2.6 Advanced Topic: Sparse Representation for Functions

In this section, we discuss sparse representation for functions.

Let us consider the function space $L^2(0,T)$, the space of all square integrable functions on $(0,T)$ in the sense of Lebesgue. That is, for any $f \in L^2(0,T)$, the L^2 norm is finite:

$$\|f\|_2 \triangleq \int_0^T |f(t)|^2 dt < \infty. \tag{2.56}$$

In this space, we can define the L^2 inner product

$$\langle f, g\rangle \triangleq \int_0^T f(t)\overline{g(t)}dt, \tag{2.57}$$

where $\overline{g(t)}$ is the complex conjugate of $g(t)$. It is well-known that under the L^2 inner product, the space L^2 becomes a Hilbert space.

Then let us consider an orthonormal basis $\{\phi_i : i \in \mathbb{Z}\}$ in $L^2(0,T)$ that satisfies

$$\langle \phi_i, \phi_j\rangle = \delta_{ij} = \begin{cases} 1, & \text{if } i = j, \\ 0, & \text{otherwise.} \end{cases} \tag{2.58}$$

Then, for any function $f \in L^2(0,T)$, there exists a complex sequence $\{\alpha_i : i \in \mathbb{Z}\}$ such that

$$f = \sum_{i=-\infty}^{\infty} \alpha_i \phi_i, \tag{2.59}$$

where the convergence is in the sense of L^2, that is,

$$\left\| f - \sum_{i=-N}^{N} \alpha_i \phi_i \right\|_2 \to 0, \tag{2.60}$$

as $N \to \infty$. The representation (2.59) is called *Fourier series*[3] of f. Given f and $\{\phi_i\}$, the coefficients are obtained by the inner product

$$\alpha_i = \langle f, \phi_i \rangle = \int_0^T f(t)\overline{\phi_i(t)}dt. \tag{2.61}$$

Exercise 2.7. Prove that (2.61) holds.

A standard basis for $L^2(0,T)$ is the *Fourier basis* defined by

$$\phi_i(t) \triangleq \frac{1}{\sqrt{T}}e^{\mathrm{j}\omega_i t}, \quad i \in \mathbb{Z}, \tag{2.62}$$

where $\mathrm{j} = \sqrt{-1}$ and $\omega_i = 2\pi i/T$. With this basis, the coefficients in (2.61) are given as

$$\alpha_i = \frac{1}{\sqrt{T}}\int_0^T f(t)\overline{e^{\mathrm{j}\omega_i t}}dt = \frac{1}{\sqrt{T}}\int_0^T f(t)e^{-\mathrm{j}\omega_i t}dt. \tag{2.63}$$

For a sufficiently smooth function, the Fourier basis gives a good solution to represent the function with a finite number of coefficients by *truncation*. That is, we approximate function f as

$$f_N = \sum_{i=-N}^{N} \alpha_i \phi_i, \quad \alpha_i = \frac{1}{\sqrt{T}}\int_0^T f(t)e^{-\mathrm{j}\omega_i t}dt. \tag{2.64}$$

Actually, this is optimal in the sense that f_N minimizes the L^2 error

$$\mathcal{E}_N(\beta_{-N}, \ldots, \beta_N) \triangleq \left\| f - \sum_{i=-N}^{N} \beta_i \phi_i \right\|_2 \tag{2.65}$$

among all coefficients $\{\beta_{-N}, \ldots, \beta_N\}$.

Now, let us consider a rectangular function on $L^2(0,1)$ defined by

$$f(t) = \begin{cases} 1, & t \in (0, 1/2), \\ -1, & t \in [1/2, 1). \end{cases} \tag{2.66}$$

Figure 2.4 (left) shows this function. We can see that this function is discontinuous. The Fourier coefficients of this function can be easily computed using (2.63). In fact, we have

$$\alpha_i = \begin{cases} -\frac{2\mathrm{j}}{\pi i}, & \text{if } \ i \text{ is odd}, \\ 0, & \text{otherwise}. \end{cases} \tag{2.67}$$

[3]This is also called as *generalized* Fourier series. Then, with the standard Fourier basis in (2.62), the series in (2.59) is called the Fourier series.

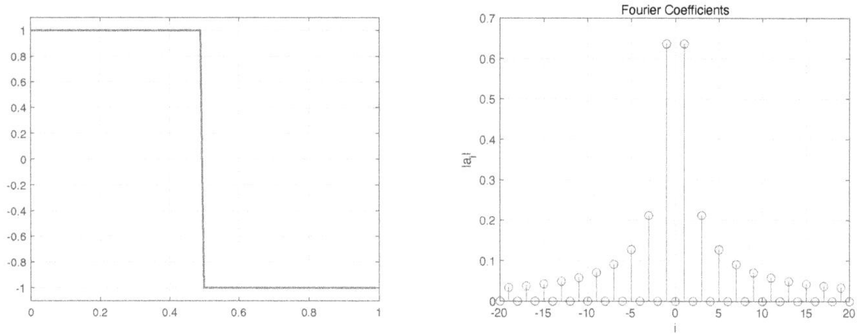

Figure 2.4: Discontinuous rectangular function $f(t)$ (left) and absolute values of its Fourier coefficients (right)

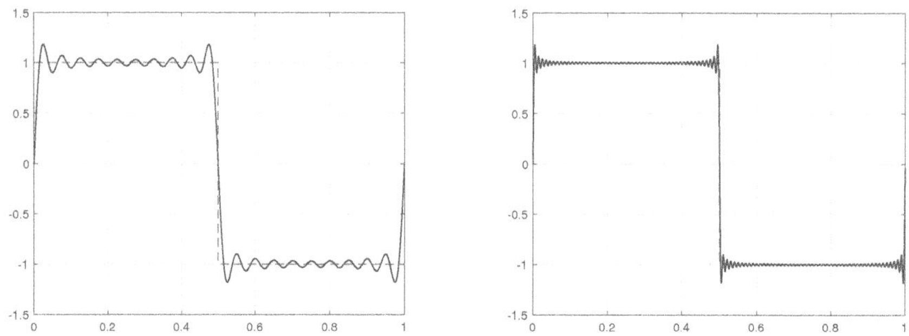

Figure 2.5: Truncated Fourier series $f_N(t)$ with $N = 20$ (left) and $N = 100$ (right)

Exercise 2.8. Show that the Fourier coefficients of the rectangular function in (2.66) are given by (2.67).

Figure 2.4 (right) shows the absolute values of the coefficients with $i = -20$ to 20. We can see that the coefficients converge to zero as i goes to $\pm\infty$. Actually, from (2.67), the coefficient sequence $\{\alpha_i\}$ converges to zero as $|i| \to \infty$ with convergence rate $O(1/i)$.

This fact suggests us to truncate the coefficients with N to obtain the approximant $f_N(t)$ in (2.64). Figure 2.5 shows the truncated Fourier series $f_N(t)$ in (2.64).

The left figure of Figure 2.5 shows $f_N(t)$ with $N = 20$. We see oscillations around the edges of the rectangular function. When we increase N to $N = 100$, we obtain another oscillative function in the right figure of Figure 2.5. This oscillation never disappears around the edges for arbitrarily large but finite N. This is called *Gibbs phenomenon.* To exactly reconstruct the

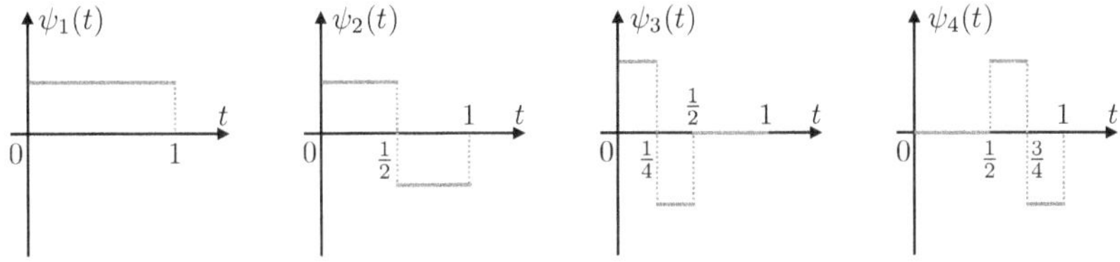

Figure 2.6: Haar functions ψ_1,ψ_2, ψ_3, and ψ_4.

shape of the rectangular function from its Fourier coefficients, you cannot truncate it but store all of the coefficients.

Let us consider another orthonormal basis in $L^2(0,1)$ called the *Haar basis* defined by the *Haar functions*

$$\psi_1(t) \triangleq 1, \tag{2.68}$$

and for $i = 2^m + k$, $k = 1, 2, \ldots, 2^m$, $m = 0, 1, 2, \ldots$,

$$\psi_i(t) \triangleq \begin{cases} \sqrt{2^m}, & t \in [2^{-m}(k-1), 2^{-m-1}(2k-1)), \\ -\sqrt{2^m}, & t \in [2^{-m-1}(2k-1), 2^{-m}k), \\ 0, & \text{otherwise}. \end{cases} \tag{2.69}$$

Figure 2.6 shows Haar functions ψ_1,ψ_2, ψ_3, and ψ_4.

Then, we can adopt a *redundant* dictionary of bases consisting of the Fourier basis in (2.62) and the Haar basis. From this dictionary, we can simply represent the rectangular function in (2.66) as

$$f(t) = \psi_2(t). \tag{2.70}$$

That is, we need to store just one coefficient under the redundant basis. This is the motivation to use a redundant dictionary and to obtain sparse representation for functions. As shown above, sparse representation of functions is achieved by sparsifying the coefficients in the Fourier series of a given function with a redundant basis dictionary.

2.7 Further Readings

The notion of redundant dictionary and sparse optimization described in this chapter is fundamental and important in this book. The redundant representation of vectors is related to *frames* and *wavelets*, for which readers can refer to nice books by Strang and Nguyen [144] and by Mallat [90]. For fundamental theory of vector spaces, called functional analysis, including norms, inner products, orthonormal bases, and Fourier series, I recommend books by Young [153] and by Yamamoto [152], which are written for scientists and engineers. For recent methods of group testing, see, for example, [1], [4].

Chapter 3

Sparse Optimization

In this chapter, we study *sparse optimization* from the viewpoint of regularization. We show some examples of curve fitting and group testing, which help readers understand the concept of sparse optimization.

> Key ideas of Chapter 3
>
> - Curve fitting and group testing can be formulated as optimization problems that choose a single solution from an underdetermined system of linear equations.
> - Regularization is used to avoid overfitting.
> - Sparse optimization is reduced to ℓ^1 optimization, which is convex and efficiently solved by numerical optimization.

3.1 Least Squares and Regularization

We begin with the least squares and regularization with simple examples.

3.1.1 Underdetermined system and minimum ℓ^2-norm solution

Let us consider the linear equation

$$\Phi \boldsymbol{x} = \boldsymbol{y}, \tag{3.1}$$

where $\boldsymbol{y} \in \mathbb{R}^m$ is a given vector, $\Phi \in \mathbb{R}^{m \times n}$ is a given matrix, and $\boldsymbol{x} \in \mathbb{R}^n$ is an unknown vector. We here assume $m < n$, that is, there are fewer equations than unknowns. Such a system of equations is said to be *underdetermied*, which appears in group testing discussed in Section 2.4 for example. We assume that Φ has full row rank, that is,

$$\operatorname{rank}(\Phi) = m. \tag{3.2}$$

Under this assumption, there exist infinitely many solutions to the equation (3.1).

Then, let us choose a single solution among the infinitely many solutions. For example, we can find the smallest ℓ^2-norm solution among them. Namely, we consider the following optimization problem:

$$\underset{\boldsymbol{x}\in\mathbb{R}^n}{\text{minimize}}\ \frac{1}{2}\|\boldsymbol{x}\|_2^2 \text{ subject to } \Phi\boldsymbol{x} = \boldsymbol{y}. \tag{3.3}$$

We call this problem the ℓ^2 *optimization problem*, and the solution the *minimum ℓ^2-norm solution.*

To solve this problem, we can use the *method of Lagrange multipliers*. First, we define the *Lagrange function*, or simply *Lagrangian*, of the optimization problem (3.3) by

$$L(\boldsymbol{x}, \boldsymbol{\lambda}) = \frac{1}{2}\boldsymbol{x}^\top\boldsymbol{x} + \boldsymbol{\lambda}^\top(\Phi\boldsymbol{x} - \boldsymbol{y}). \tag{3.4}$$

The variable $\boldsymbol{\lambda} \in \mathbb{R}^m$ is called the *Lagrange multiplier.*

Then, we can obtain the optimal solution to (3.3) by finding the stationary point $(\boldsymbol{x}^*, \boldsymbol{\lambda}^*)$ of the Lagrange function L. By differentiating L by the variable $\boldsymbol{x}$, we have

$$\frac{\partial L}{\partial \boldsymbol{x}} = \frac{\partial}{\partial \boldsymbol{x}}\left(\frac{1}{2}\boldsymbol{x}^\top\boldsymbol{x} + \boldsymbol{\lambda}^\top\Phi\boldsymbol{x}\right) = \boldsymbol{x} + \Phi^\top\boldsymbol{\lambda}. \tag{3.5}$$

It follows that the stationary point $(\boldsymbol{x}^*, \boldsymbol{\lambda}^*)$ satisfies

$$\boldsymbol{x}^* + \Phi^\top\boldsymbol{\lambda}^* = \boldsymbol{0}. \tag{3.6}$$

Then differentiating L by $\boldsymbol{\lambda}$ gives

$$\frac{\partial L}{\partial \boldsymbol{\lambda}} = \Phi\boldsymbol{x} - \boldsymbol{y}, \tag{3.7}$$

and hence

$$\Phi\boldsymbol{x}^* - \boldsymbol{y} = \boldsymbol{0}. \tag{3.8}$$

From this and (3.6), we have

$$-\Phi\Phi^\top\boldsymbol{\lambda}^* = \boldsymbol{y}. \tag{3.9}$$

Since Φ has full row rank, the matrix $\Phi\Phi^\top$ is non-singular and has its inverse. Therefore, from (3.9) we have

$$\boldsymbol{\lambda}^* = -(\Phi\Phi^\top)^{-1}\boldsymbol{y}. \tag{3.10}$$

Assigning this to (3.6) gives the minimum ℓ^2-norm solution $\boldsymbol{x}^*$ as

$$\boldsymbol{x}^* = \Phi^\top(\Phi\Phi^\top)^{-1}\boldsymbol{y}. \tag{3.11}$$

In summary, if we are given a full-row-rank matrix Φ and a vector $\boldsymbol{y}$, we can compute the minimum ℓ^2-norm solution by the formula (3.11).

Exercise 3.1. Find the minimum ℓ^2-norm solution to the following equation with unknowns x_1 and x_2:

$$a_1 x_1 + a_2 x_2 = 1, \tag{3.12}$$

where a_1 and a_2 are nonzero real numbers.

Exercise 3.2. Let $\Phi \in \mathbb{R}^{m\times n}$. Prove that $\Phi\Phi^\top$ is invertible if Φ has full row rank.

3.1.2 Regression and least squares

Here we consider the curve fitting problem as an example of least squares. Suppose we are given the following two-dimensional dataset

$$\mathcal{D} = \{(t_1, y_1), (t_2, y_2), \dots, (t_m, y_m)\}. \tag{3.13}$$

Let us consider a polynomial of order $n-1$,

$$y = f(t) = a_{n-1}t^{n-1} + a_{n-2}t^{n-2} + \cdots + a_1 t + a_0. \tag{3.14}$$

Polynomial curve fitting is to find coefficients $a_0, a_1, \dots, a_{n-1}$ with which the polynomial curve has the best fit to the m-point data (see Figure 3.1 for example). For example, $t_1, t_2, \dots, t_m$ are sampling instants, and $y_1, y_2, \dots, y_m$ are temperature data from a sensor at a position. From these data, we often want to know the curve behind the data. We call such data analysis the *regression analysis*.

First, we consider an *interpolating polynomial* that interpolates the given data as shown in Figure 3.1. The polynomial curve (3.14) goes through the data points (3.13), and hence we have m linear equations with

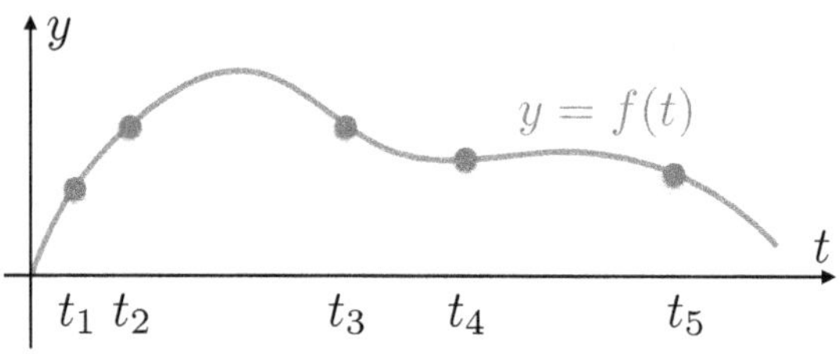

Figure 3.1: Interpolating polynomial

unknowns a_{n-1}, a_{n-2}, …, a_1, a_0:

$$\begin{aligned} a_{n-1}t_1^{n-1} + a_{n-2}t_1^{n-2} + \cdots + a_1 t_1 + a_0 &= y_1, \\ a_{n-1}t_2^{n-1} + a_{n-2}t_2^{n-2} + \cdots + a_1 t_2 + a_0 &= y_2, \\ &\cdots \\ a_{n-1}t_m^{n-1} + a_{n-2}t_m^{n-2} + \cdots + a_1 t_m + a_0 &= y_m. \end{aligned} \tag{3.15}$$

Define a matrix

$$\Phi \triangleq \begin{bmatrix} t_1^{n-1} & t_1^{n-2} & \dots & t_1 & 1 \\ t_2^{n-1} & t_2^{n-2} & \dots & t_2 & 1 \\ \vdots & \vdots & \ddots & \vdots & \\ t_m^{n-1} & t_m^{n-2} & \dots & t_m & 1 \end{bmatrix} \in \mathbb{R}^{m \times n}, \tag{3.16}$$

and vectors

$$\boldsymbol{x} \triangleq \begin{bmatrix} a_{n-1} \\ a_{n-2} \\ \vdots \\ a_1 \\ a_0 \end{bmatrix} \in \mathbb{R}^n, \quad \boldsymbol{y} \triangleq \begin{bmatrix} y_1 \\ y_2 \\ \vdots \\ y_m \end{bmatrix} \in \mathbb{R}^m. \tag{3.17}$$

Then the system of linear equations (3.15) can be represented in a matrix form $\Phi\boldsymbol{x} = \boldsymbol{y}$. The matrix Φ is known as a *Vandermonde matrix*, and if $m = n$, then Φ is a square matrix and its determinant is given by

$$\det(\Phi) = \prod_{1 \le i < j \le m} (t_i - t_j) = (t_1 - t_2)(t_1 - t_3) \cdots (t_{m-1} - t_m). \tag{3.18}$$

It follows that if

$$t_i \neq t_j \quad \text{for all} \quad i, j \quad \text{such that} \quad i \neq j, \tag{3.19}$$

then Φ is non-singular and has its inverse. Hence, the solution $\boldsymbol{x}^*$ to equation (3.15) is given by using Φ^{-1} as

$$\boldsymbol{x}^* = \Phi^{-1}\boldsymbol{y}. \tag{3.20}$$

In summary, if one chooses a polynomial of order $m-1$ for m data points that satisfy (3.19), then the coefficients of the interpolating polynomial can be uniquely obtained by the formula (3.20).

Example 3.1. Let us consider the following data.

t	1	2	3	...	14	15
y	2	4	6	...	28	30

The data are obtained from a linear relation $y = 2t$. By using these 15 data points, we find a 14th-order interpolating polynomial. Now, we use a useful computational software, Python, to compute the matrix inversion in (3.20). The following is a program to obtain the coefficients.

Program 3.1: Python program for the coefficients of the interpolating polynomial.

```
import numpy as np

# Data
t = np.arange(1, 16)
y = 2 * t

# Vandermonde matrix
Phi = np.vander(t)

# Coefficients of interpolating polynomial
x = np.linalg.inv(Phi) @ y
print(x)
```

In this program, `numpy` is a library that underpins much of scientific computing in Python, which provides efficient and convenient ways to handle numerical data. The command `vander` is a function in the `numpy` library to compute the Vandermonde matrix in (3.16). The inverse of Φ is computed by the `inv` function in the `numpy.linalg` module, and the multiplication $\Phi^{-1}\boldsymbol{y}$ is executed by the `@` operator.

Running this program, we obtain

$$x = \begin{bmatrix} 2.85231728 \times 10^{-19} \\ -1.49471669 \times 10^{-17} \\ 3.13375085 \times 10^{-16} \\ -4.28216490 \times 10^{-15} \\ 6.10900219 \times 10^{-14} \\ -8.59645688 \times 10^{-13} \\ 8.46611670 \times 10^{-12} \\ -4.87716534 \times 10^{-11} \\ 1.73088210 \times 10^{-10} \\ -3.61524144 \times 10^{-10} \\ 4.81122697 \times 10^{-10} \\ -3.00133252 \times 10^{-10} \\ 3.69254849 \times 10^{-10} \\ 2.00000000 \\ 3.18323146 \times 10^{-11} \end{bmatrix}. \tag{3.21}$$

This result shows that the second value from the bottom is 2, and the other values are almost zero. That is, the coefficients are given by $a_1 = 2$ and $a_i = 0$ for $i \neq 1$, and the interpolating polynomial is $y = 2t$. This is the right solution. □

Real data include *noise*. Let us add Gaussian noise with zero mean and variance 0.5^2 to the data y in Example 3.1, and find the interpolating polynomial. The obtained curve is shown in Figure 3.2. The interpolating polynomial exactly goes through the data points, but the curve is significantly affected by noise, and is very different from the original relationship $y = 2t$. Such a phenomenon is called *overfitting*.

The reason for overfitting is that the order of the polynomial is too high. If we previously know that the original curve is of first order, then we can assume a first-order polynomial (i.e. a line) $y = a_1 t + a_0$, and find the coefficients a_0 and a_1 with which the line has the best fit to the data. If the data is noisy, it is obviously impossible to obtain a line that goes through all the data points. However, it is not a problem at all if the line does not interpolate the noisy data.

Now, let us reformulate our problem of curve fitting for noisy data. We measure the distance between the polynomial and the data points by the ℓ^2 norm (Euclidean norm). For noisy data, we do not require the curve to go through the data points since it is in general impossible. We find

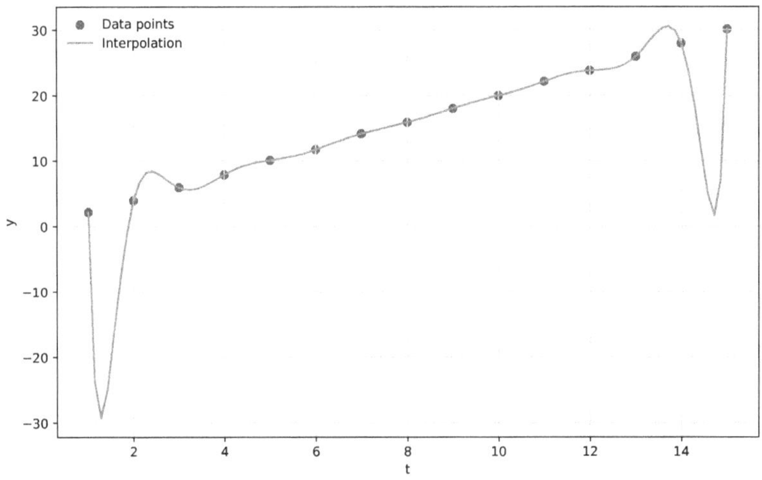

Figure 3.2: 14th-order interpolating polynomial with noisy data

the curve that is as close to the data points as possible. The optimization problem is described as follows:

$$\underset{\boldsymbol{x}\in\mathbb{R}^n}{\text{minimize}}\ \frac{1}{2}\|\Phi\boldsymbol{x}-\boldsymbol{y}\|_2^2, \tag{3.22}$$

where $\Phi \in \mathbb{R}^{m\times n}$ is the Vandermonde matrix defined in (3.16). We call the optimization in (3.22) the *least squares*. If we assume $n < m$, that is, if the order of the polynomial is less than $m-2$, then the number of unknowns is less than that of equations. In this case, the matrix Φ is a *tall matrix*, and the equation $\Phi\boldsymbol{x} = \boldsymbol{y}$ has no solution in general. If the condition (3.19) holds, it is easily shown that the solution to (3.22) uniquely exists. In fact, if (3.19) holds, then the Vandermonde matrix Φ has *full column rank*. Note that a matrix $\Phi \in \mathbb{R}^{m\times n}$ is said to have full column rank if the n column vectors in Φ are linearly independent. In other words, $\Phi \in \mathbb{R}^{m\times n}$ has full column rank if and only if Φ is *injective*, or

$$\text{rank}(\Phi) = n. \tag{3.23}$$

Then, the unique solution to the optimization problem in (3.22) is given by

$$\boldsymbol{x}^* = (\Phi^\top\Phi)^{-1}\Phi^\top\boldsymbol{y}. \tag{3.24}$$

We call this the *least squares solution*. As the minimum ℓ^2-norm solution in (3.11), the least squares solution is also given in a closed form.

Exercise 3.3. Prove that the solution to the optimization problem (3.22) is given by (3.24).

Exercise 3.4. Let ϕ_i denote the i-th column vector in matrix $\Phi \in \mathbb{R}^{m \times n}$, that is,

$$\Phi = \begin{bmatrix} \phi_1 & \phi_2 & \dots & \phi_n \end{bmatrix}. \tag{3.25}$$

Then define the *residual* between the data $\boldsymbol{y}$ and the optimal estimation $\Phi \boldsymbol{x}^*$ with (3.24) by

$$\boldsymbol{r} \triangleq \boldsymbol{y} - \Phi \boldsymbol{x}^*. \tag{3.26}$$

Prove that the residual satisfies

$$\langle \boldsymbol{\phi}_i, \boldsymbol{r} \rangle = 0, \quad \forall i \in \{1, 2, \dots, n\}. \tag{3.27}$$

Also, by using this fact, show that the residual $\boldsymbol{r}$ is orthogonal to $\Phi \boldsymbol{x}^*$.

Example 3.2. Let us consider Example 3.1 with additive Gaussian noise with zero mean and variance 0.5^2. We assume the curve is a first-order polynomial modeled by $y = a_1 t + a_0$. A Python program to obtain the least squares solution to this problem is given as follows:

Program 3.2: Python program for least squares solution.

```
import numpy as np

# Data
np.random.seed(1)  # random seed
t = np.arange(1, 16)
y = 2 * t + np.random.randn(15) * 0.5

# Vandermonde matrix
Phi15 = np.vander(t, 15)
Phi = Phi15[:, 13:15]

# Least squares solution
Phi_transpose = Phi.T
x = np.linalg.inv(Phi_transpose @ Phi) @
    Phi_transpose @ y
print(x)
```

In this program, `randn(15)` is a function in the `numpy.random` module that returns normally distributed random numbers (i.e., Gaussian noise)

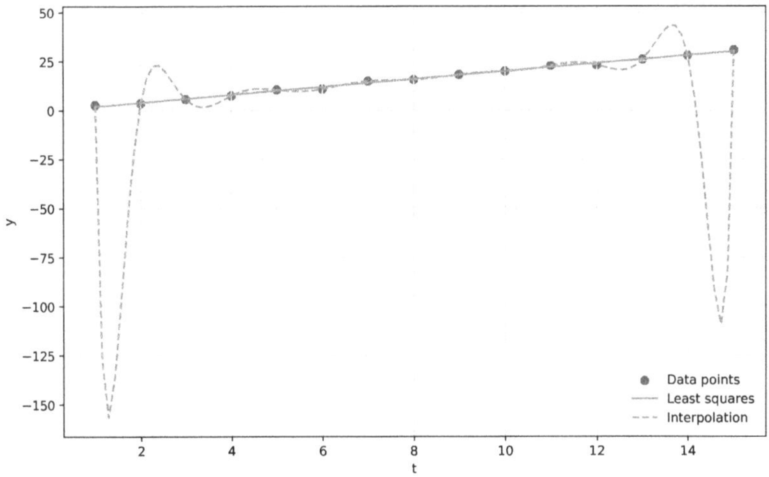

Figure 3.3: Least square solution (solid line) and 14th-order interpolating polynomial (dashed curve)

with zero mean and variance 1 of size 15. The matrix variable `Phi15` is the Vandermonde matrix of size 15×15, and in the 10th line we extract the 14th and 15th columns, which are related to coefficients a_1 and a_0, to make matrix `Phi` of size 15×2. The result is shown below.

$$\boldsymbol{x} = \begin{bmatrix} 1.99907309 \\ -0.03065618 \end{bmatrix}. \tag{3.28}$$

Figure 3.3 shows the line $y = a_1 t + a_0$ with these coefficients. While the 14th-order interpolating polynomial implies overfitting, the least squares line shows a good result. □

3.1.3 Ridge regularization

As we have discussed in the previous section, we can avoid overfitting by the least squares method with an appropriate order of the polynomial, which is less than the number of data points. However, what can we do if we do not know the proper order in advance? In this case, we can adopt *regularization*. Let us begin with a simple example.

Example 3.3. Suppose that we are given the following dataset

$$\mathcal{D} = \{(t_1, y_1), (t_2, y_2), \ldots, (t_m, y_m)\}, \tag{3.29}$$

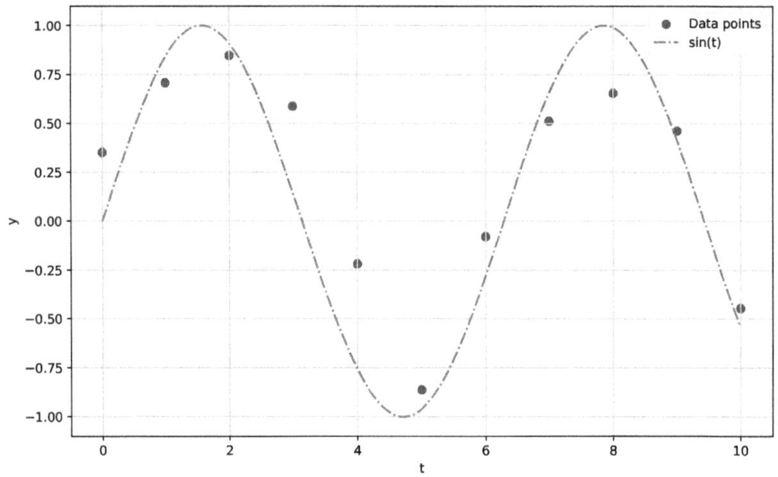

Figure 3.4: 11 data points from a sinusoid (dashed curve)

which is generated from a sinusoid $y = \sin(t)$. We consider sampling instants

$$t_1 = 0,\ t_2 = 1,\ t_3 = 2, \ldots, t_{11} = 10, \tag{3.30}$$

on which the output points $y_1, y_2, \ldots, y_{11}$ are obtained as

$$y_i = \sin(t_i) + \epsilon_i, \quad i = 1, 2, \ldots, 11, \tag{3.31}$$

where ϵ_i is Gaussian noise with zero mean and variance 0.2^2 added at time t_i independently. The following table shows the obtained data.

t_i	0	1	2	3	4	5
y_i	0.3528	0.7076	0.8473	0.5879	-0.2187	-0.8624
t_i	6	7	8	9	10	
y_i	-0.0789	0.5122	0.6549	0.4622	-0.4460	

Figure 3.4 shows the data points and the original sinusoidal curve.

For these data, we first find a 10th-order polynomial that interpolates the data points by using (3.20). Figure 3.5 shows the result. Affected by the noise, the curve is very oscillative and shows overfitting. We then take a 6th-order polynomial and compute the least squares solution by (3.24). Figure 3.6 shows the result. From this figure, we have a better fit than the interpolating function shown in Figure 3.5. □

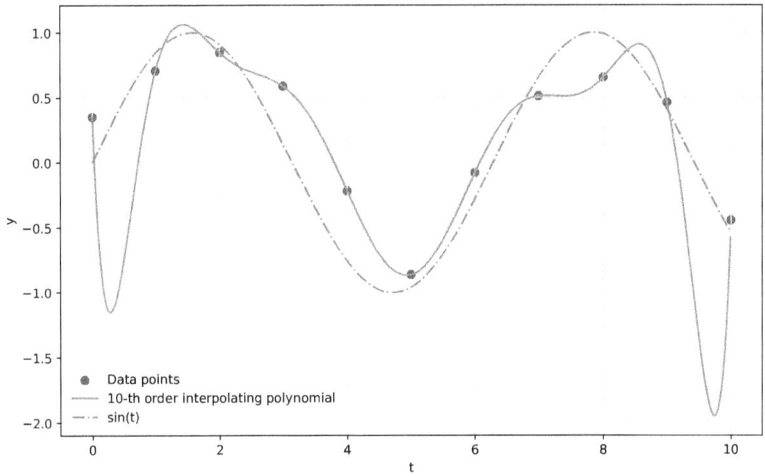

Figure 3.5: 10th-order interpolating polynomial (solid curve) and the original sinusoid (dashed curve)

In the above example, the order 6 was chosen by computing curves of all orders from 1 to 10, and comparing the reconstructed curve with the original sinusoid. However, this can be done *if we previously know the original sinusoid.* This is impossible in real applications. That is, we do not know the optimal polynomial order from data in advance[1].

To see the difference between the 10th-order interpolating polynomial and the 6th-order least squares polynomial, we compare their coefficients. Let us denote by $\boldsymbol{x}_{10}$ and $\boldsymbol{x}_6$ respectively the 10th and 6th-order polyno-

[1]For choosing the polynomial order, we can adopt the method of *cross validation*, dividing the data into two sets; training and test data. This is actually a powerful method but it takes a lot of time to perform. See [10] for details.

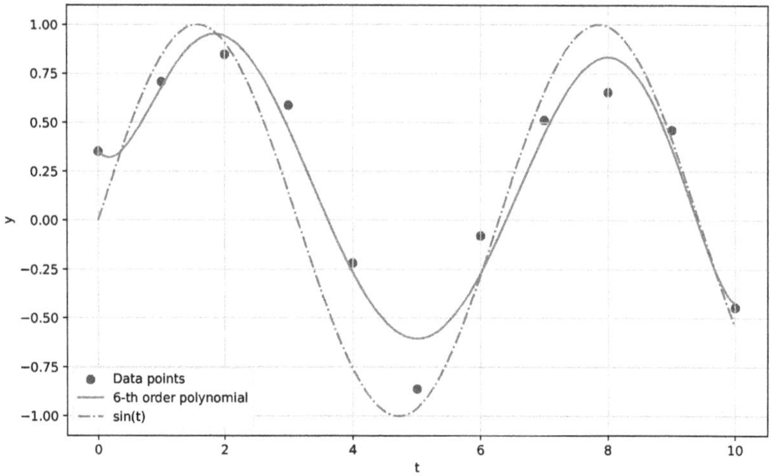

Figure 3.6: Least squares solution with 6th-order polynomial (solid curve) and the original sinusoid (dashed curve)

mials. They are obtained as

$$
\boldsymbol{x}_{10} = \begin{bmatrix} 0.3528 \\ -13.0687 \\ \mathbf{37.3672} \\ \mathbf{-41.5757} \\ \mathbf{24.7010} \\ -8.7587 \\ 1.9368 \\ -0.2690 \\ 0.0228 \\ -0.0011 \\ 0.0000 \end{bmatrix}, \quad \boldsymbol{x}_6 = \begin{bmatrix} 0.3549 \\ \mathbf{-0.3993} \\ \mathbf{1.3586} \\ \mathbf{-0.7883} \\ 0.1704 \\ -0.0157 \\ 0.0005 \end{bmatrix}, \tag{3.32}
$$

where the boldface numbers are the largest three elements in their absolute values. We can observe that the boldface values in $\boldsymbol{x}_{10}$ are much larger than those in $\boldsymbol{x}_6$. This is a cause of oscillation in the 10th-order interpolating curve.

From the above observation, we try to minimize both the squared error $\|\Phi\boldsymbol{x} - \boldsymbol{y}\|_2^2$ and the squared ℓ^2 norm $\|\boldsymbol{x}\|_2^2$ of the coefficient vector $\boldsymbol{x}$ at the same time. This is formulated as the following optimization problem:

$$\underset{\boldsymbol{x}\in\mathbb{R}^n}{\text{minimize}}\ \frac{1}{2}\|\Phi\boldsymbol{x}-\boldsymbol{y}\|_2^2+\frac{\lambda}{2}\|\boldsymbol{x}\|_2^2. \tag{3.33}$$

We call this optimization the *regularized least squares*, or *ridge regression*. The additional term $\frac{\lambda}{2}\|\boldsymbol{x}\|_2^2$ is called the *regularization term*, and the parameter $\lambda > 0$ the *regularization parameter*, which controls the balance between the error in curve fitting and the ℓ^2 norm of the coefficients.

As in the least squares solution, the solution to the regularized least squares in (3.33) can be obtained in a closed form:

$$\boldsymbol{x}^* = (\lambda I + \Phi^\top\Phi)^{-1}\Phi^\top\boldsymbol{y}. \tag{3.34}$$

Exercise 3.5. Prove that the solution to (3.33) is given by (3.34).

Example 3.4. Here we consider an example of regularization. With the data given in Example 3.3, we compute a 10th-order polynomial by the regularized least squares. We take the regularization parameter $\lambda = 1$, and compute the solution $\boldsymbol{x}^*$ by the formula (3.34). The obtained coefficients are as follows:

$$\boldsymbol{x}^* = \begin{bmatrix} \mathbf{0.2435} \\ 0.1810 \\ \mathbf{0.2134} \\ 0.1092 \\ \mathbf{-0.1821} \\ 0.0607 \\ -0.0087 \\ 0.0006 \\ -0.0000 \\ 0.0000 \\ 0.0000 \end{bmatrix}. \tag{3.35}$$

The boldface values are the three largest elements in the absolute values. Compared with the coefficients $\boldsymbol{x}_{10}$ in (3.32) of the 10th-order interpolating polynomial, the values in $\boldsymbol{x}^*$ are much smaller. The curve with the regularized least squares is shown in Figure 3.7. We can see that the 10th-order polynomial by the regularized least squares shows comparable accuracy to the 6th-order least squares solution. □

3.1.4 Weighted ridge regression

Here we further consider the problem of polynomial interpolation. In the regularized least squares, we minimize the cost function in (3.33) with the

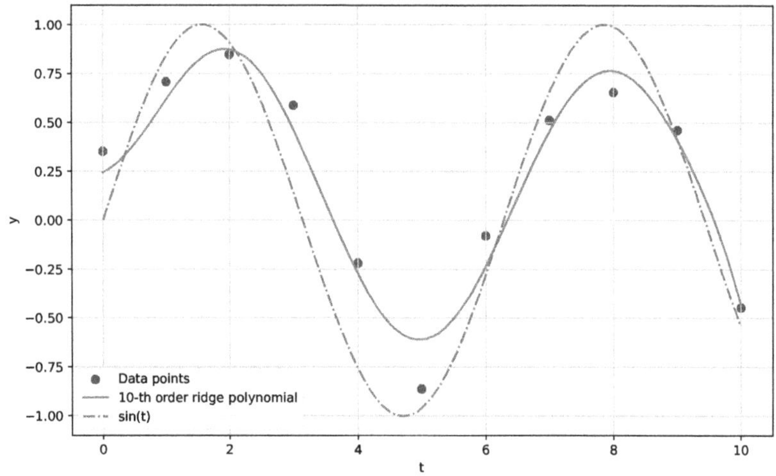

Figure 3.7: Regularized least squares solution with 10th-order polynomial (solid curve) and the original sinusoid (dashed curve)

regularization term $\|\boldsymbol{x}\|_2^2$. This is to make the coefficient vector $\boldsymbol{x}$ not so large. Instead of this, we consider the L^2 norm of the polynomial $f(t)$. The L^2 norm of a function $f(t)$, $t \in [t_1, t_m]$ is defined as

$$\|f\|_{L^2} \triangleq \sqrt{\int_{t_1}^{t_m} |f(t)|^2 \, dt}. \tag{3.36}$$

Since $f(t)$ is a polynomial, namely,

$$f(t) = \sum_{i=0}^{n-1} a_i t^i, \tag{3.37}$$

its L^2 norm can be computed as

$$\begin{aligned} \|f\|_{L^2}^2 &= \int_{t_1}^{t_m} \left(\sum_{i=0}^{n-1} a_i t^i\right)\left(\sum_{j=0}^{n-1} a_j t^j\right) dt \\ &= \sum_{i=0}^{n-1}\sum_{j=0}^{n-1} a_i a_j \int_{t_1}^{t_m} t^{i+j} \, dt \\ &= \sum_{i=0}^{n-1}\sum_{j=0}^{n-1} a_i a_j \frac{t_m^{i+j+1} - t_1^{i+j+1}}{i+j+1} \\ &= \boldsymbol{x}^\top Q \boldsymbol{x}, \end{aligned} \tag{3.38}$$

Table 3.1: Summary of optimization problems with ℓ^2 norm

Problem	Size	Problem	Solution
min ℓ^2 norm	$m < n$	$\min_{x} \frac{1}{2}\|x\|_2^2$ s.t. $\Phi x = y$	$\Phi^\top(\Phi\Phi^\top)^{-1}y$
least squares	$m > n$	$\min_{x} \frac{1}{2}\|\Phi x - y\|_2^2$	$(\Phi^\top\Phi)^{-1}\Phi^\top y$
ridge regression	any	$\min_{x} \frac{1}{2}\|\Phi x - y\|_2^2 + \frac{\lambda}{2}\|x\|_2^2$	$(\lambda I + \Phi^\top\Phi)^{-1}\Phi^\top y$
weighted ridge regression	any	$\min_{x} \frac{1}{2}\|\Phi x - y\|_2^2 + \frac{\lambda}{2}\|\Psi x\|_2^2$	$(\lambda\Psi^\top\Psi + \Phi^\top\Phi)^{-1}\Phi^\top y$

where $Q = [Q_{ij}]$ is a matrix defined by

$$Q_{ij} = \frac{t_m^{i+j+1} - t_1^{i+j+1}}{i+j+1}, \qquad i, j = 0, 1, \ldots, n-1. \tag{3.39}$$

Now, from the definition of the L^2 norm, we have $\|f\|_{L^2} \geq 0$ for any polynomial f, and $\|f\|_{L^2} = 0$ if and only if $f = 0$. This means that for any $\boldsymbol{x} \in \mathbb{R}^n$, we have $\boldsymbol{x}^\top Q\boldsymbol{x} \geq 0$ and $\boldsymbol{x}^\top Q\boldsymbol{x} = 0$ if and only if $\boldsymbol{x} = \mathbf{0}$. That is, the matrix Q is *positive definite.*

Now, we consider a regularization problem minimizing

$$\frac{1}{2}\|\Phi\boldsymbol{x} - \boldsymbol{y}\|_2^2 + \frac{\lambda}{2}\|f\|_{L^2}^2 = \frac{1}{2}\|\Phi\boldsymbol{x} - \boldsymbol{y}\|_2^2 + \frac{\lambda}{2}\|\Psi\boldsymbol{x}\|_2^2, \tag{3.40}$$

where Ψ is a matrix that satisfies $Q = \Psi^\top\Psi$. This is called the *weighted ridge regression.* The solution is obtained by

$$\boldsymbol{x}^* = (\lambda\Psi^\top\Psi + \Phi^\top\Phi)^{-1}\Phi^\top\boldsymbol{y}. \tag{3.41}$$

Exercise 3.6. Prove that (3.41) is the solution to the optimization problem minimizing (3.40).

3.1.5 Summary of ℓ^2-norm optimization

Now we summarize the curve fitting problem by a polynomial of order $m-1$:

$$y = f(t) = a_{n-1}t^{n-1} + a_{n-2}t^{n-2} + \cdots + a_1 t + a_0, \tag{3.42}$$

with the following dataset:

$$\mathcal{D} = \{(t_1, y_1), (t_2, y_2), \ldots, (t_m, y_m)\}. \tag{3.43}$$

Table 3.1 shows the summary.

Exercise 3.7. Prove that for any matrix $\Phi \in \mathbb{R}^{m\times n}$ and any number $\lambda > 0$, matrices $\lambda I + \Phi\Phi^\top$ and $\lambda I + \Phi^\top\Phi$ are invertible and satisfy

$$\Phi^\top(\lambda I + \Phi\Phi^\top)^{-1} = (\lambda I + \Phi^\top\Phi)^{-1}\Phi^\top. \tag{3.44}$$

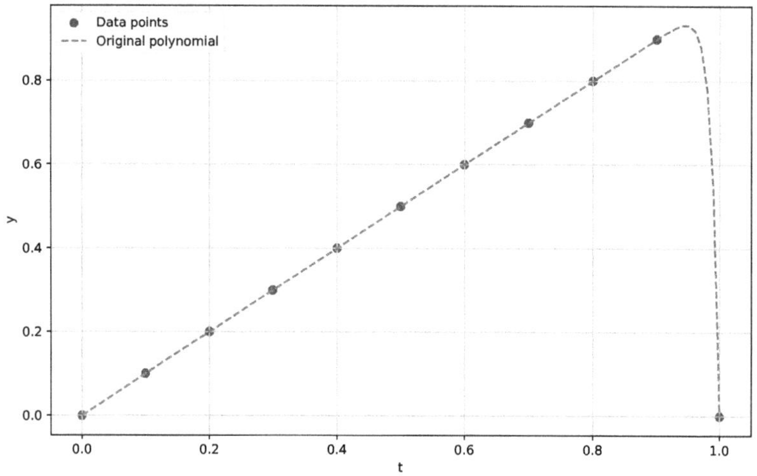

Figure 3.8: Sparse polynomial $y = -t^{80} + t$ and sampled data

3.2 Sparse Polynomial and ℓ^1-norm Optimization

Here we consider yet another example of curve fitting. Let us consider an 80th-order polynomial

$$y = -t^{80} + t. \tag{3.45}$$

From this polynomial, we sample data points with sampling instants

$$t_1 = 0, t_2 = 0.1, t_3 = 0.2, \ldots, t_{11} = 1, \tag{3.46}$$

to obtain

$$\mathcal{D} = \{(t_1, y_1), (t_2, y_2), \ldots, (t_{11}, y_{11})\}, \quad y_i = -t_i^{80} + t_i. \tag{3.47}$$

Figure 3.8 shows the curve of the 80th-order polynomial in (3.45) and the generated data in (3.47).

We assume that the order of the original polynomial is previously known to be at most 80. Then, can we reconstruct the original curve in (3.45) from the dataset $\mathcal{D}$? In this case, there are infinitely many interpolating polynomials with order at most 80 that go through all the data points. In fact, the Vandermonde matrix Φ in (3.16) is a fat matrix of size 11×81, which has full row rank since the condition (3.19) holds. Therefore, there exist infinitely many solutions to the linear equation $\Phi \boldsymbol{x} = \boldsymbol{y}$, where $\boldsymbol{x}$ is a column vector consisting of 81 unknown coefficients, and $\boldsymbol{y}$ is a column vector consisting of data $y_1, y_2, \ldots, y_{11}$. As mentioned in Section 3.1, we need additional *proofs* to obtain the unique solution.

Let us look again at the original 80th-order polynomial in (3.45). The coefficients of this polynomial are all zero but two coefficients. In other words, the coefficient vector $\boldsymbol{x} = (a_{80}, a_{79}, \ldots, a_0)$ is *sparse*, that is, the ℓ^0 norm of $\boldsymbol{x}$ is sufficiently small. We call such a polynomial a *sparse polynomial*. We assume that the following fact can be additionally used as our *proof*.

The original polynomial is sparse.

Note that we can use the sparsity property of the original polynomial but the number of non-zero coefficients (i.e., $\|\boldsymbol{x}\|_0$) is assumed to be unknown.

Borrowing the idea of the optimization mentioned in Section 2.3, we use the ℓ^0 norm as the cost function, and consider the following optimization problem:

$$\underset{\boldsymbol{x}\in\mathbb{R}^n}{\text{minimize}} \ \|\boldsymbol{x}\|_0 \quad \text{subject to} \quad \Phi\boldsymbol{x} = \boldsymbol{y}. \tag{3.48}$$

As mentioned in Section 2.5, this is quite hard to solve using the exhaustive search method when the problem size is large.

The key idea of sparse optimization is to use the ℓ^1 *norm*

$$\|\boldsymbol{x}\|_1 = \sum_{i=1}^{n} |x_i|, \tag{3.49}$$

instead of the ℓ^0 norm. That is, we consider the following optimization problem as a relaxation of the ℓ^0 optimization (3.48):

$$\underset{\boldsymbol{x}\in\mathbb{R}^n}{\text{minimize}} \ \|\boldsymbol{x}\|_1 \quad \text{subject to} \quad \Phi\boldsymbol{x} = \boldsymbol{y}. \tag{3.50}$$

We call this optimization the ℓ^1 *optimization*. The method to obtain a sparse vector by the ℓ^1 optimization is known as the *basis pursuit*.

The ℓ^1 optimization problem in (3.50) is to find the smallest ℓ^1-norm vector on a hyperplane $\{\boldsymbol{x} \in \mathbb{R}^n : \Phi\boldsymbol{x} = \boldsymbol{y}\}$. As illustrated in Figure 2.2 (p. 22), the contour of the ℓ^1 norm ($\|\boldsymbol{x}\|_1 = c$) is a diamond whose corners are on the axes. The optimal solution to (3.50) is obtained (in the 2-dimensional case) by enlarging the contour $\|\boldsymbol{x}\|_1 = c$ from $c = 0$ until the contour touches the line $\{\boldsymbol{x} \in \mathbb{R}^2 : \Phi\boldsymbol{x} = \boldsymbol{y}\}$. As shown in Figure 3.9, the ℓ^1 contour touches the line almost surely at one of the corners. Since each corner is on one of the axes, the optimal solution satisfies $\|\boldsymbol{x}^*\|_0 = 1 < 2$, and hence it is sparse. This is an intuitive explanation of why minimizing ℓ^1 norm gives a sparse solution.

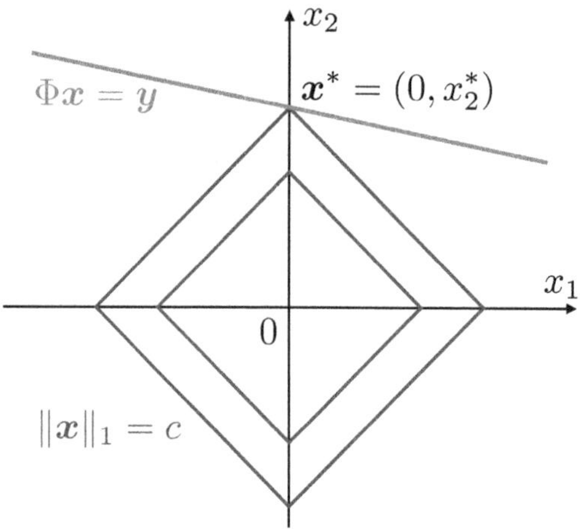

Figure 3.9: ℓ^1 optimization in $\mathbb{R}^2$: the contour $\{\boldsymbol{x} : \|\boldsymbol{x}\|_1 = c\}$ touches the line $\{\boldsymbol{x} : \Phi\boldsymbol{x} = \boldsymbol{y}\}$ at one of the corners that are on axes.

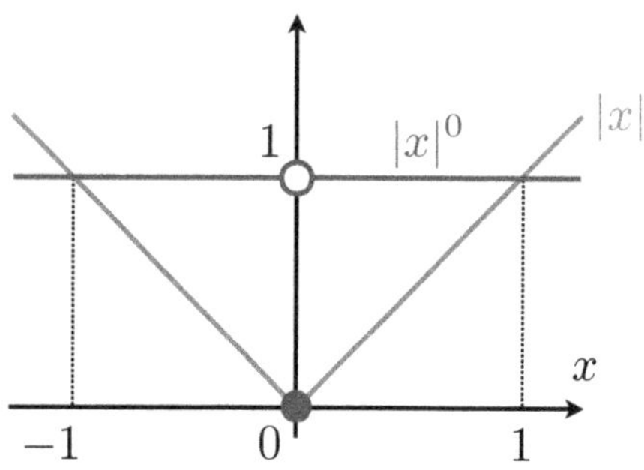

Figure 3.10: Relation between $|x|$ and $|x|^0$.

The relation between the ℓ^0 and ℓ^1 norms is intuitively understood as follows. By definition, the ℓ^0 norm can be rewritten as

$$\|\boldsymbol{x}\|_0 = \sum_{i=1}^{n} |x_i|^0, \tag{3.51}$$

where $|x|^0 \triangleq 1$ if $x \neq 0$ and $0^0 \triangleq 0$. Figure 3.10 shows the graph of $|x|^0$. From this figure, it is easily seen that the ℓ^0 norm is non-convex. On the other hand, the ℓ^1 norm

$$\|\boldsymbol{x}\|_1 = \sum_{i=1}^{n} |x_i|, \tag{3.52}$$

is the sum of absolute values $|x_i|$, which is convex as shown in Figure 3.10. The ℓ^1 norm is the best convex approximation of the ℓ^0 norm in the sense that it has the minimum exponent $p = 1$ among ℓ^p norms that are convex.

Theoretically, the ℓ^1 norm is also understood as the *convex relaxation* of the ℓ^0 norm. That is, the ℓ^1 norm is the second conjugate $\|\cdot\|_0^{**}$ of $\|\cdot\|_0$. See [149, Section 1.3] for details.

In the ℓ^1 optimization problem in (3.50), the cost function $\|\boldsymbol{x}\|_1$ is a convex function of $\boldsymbol{x}$, and the constraint set $\{\boldsymbol{x} \in \mathbb{R}^n : \Phi\boldsymbol{x} = \boldsymbol{y}\}$ is a convex set. Therefore, the problem is a *convex optimization problem*[2], for which *numerical optimization* by using a computer can give a numerical solution much faster than the exhaustive search for the ℓ^0 optimization in (3.48).

To obtain a sparse vector, we can also use the idea of regularization for sparse optimization. In the case of noisy data, we can formulate the curve fitting as the ℓ^0 *regularization* described below:

$$\underset{\boldsymbol{x}\in\mathbb{R}^n}{\text{minimize}} \ \frac{1}{2}\|\Phi\boldsymbol{x} - \boldsymbol{y}\|_2^2 + \lambda\|\boldsymbol{x}\|_0. \tag{3.53}$$

Unfortunately, this optimization is also a combinatorial problem, and hard to solve if the problem size is large. Instead of the ℓ^0 norm for the regularization term, we use the ℓ^1 norm and consider the following optimization problem:

$$\underset{\boldsymbol{x}\in\mathbb{R}^n}{\text{minimize}} \ \frac{1}{2}\|\Phi\boldsymbol{x} - \boldsymbol{y}\|_2^2 + \lambda\|\boldsymbol{x}\|_1. \tag{3.54}$$

This is called ℓ^1 *regularization*, or *LASSO* (Least Absolute Shrinkage and Selection Operator). The cost function in (3.54) is a convex function of $\boldsymbol{x}$, and hence the optimization is a convex optimization problem, which can also be solved very efficiently.

In this section, we have introduced the important idea to approximate the ℓ^0 norm, which is non-convex and discontinuous, by the ℓ^1 norm, which is convex. A question is when the convex optimization problem (3.50), or (3.54) can give the solution to the original ℓ^0 optimization (3.48), or (3.53). Very interestingly, in many applications (e.g., signal/image processing), the solution to the ℓ^1 norm optimization is equivalent to (or sufficiently close to) the ℓ^0-norm solution. In fact, there exist many theorems for the equivalence between ℓ^0 and ℓ^1 optimizations. From these facts, ℓ^1 optimization is often said to be sparse optimization. For seeking sparse solutions, there also exist many methods other than ℓ^1 optimization, for example, greedy methods, which we will see in detail in Chapter 5, or ℓ^p-norm optimization with $p \in (0, 1)$. In the next section, we will show how to solve the ℓ^1 optimization (3.50) or ℓ^1 regularization (3.54) by using Python.

[2]For the mathematical definition of convexity and convex optimization, see Chapter 4.

3.3 Python Examples

3.3.1 Sparse polynomial curve fitting

The optimization problems (3.50) and (3.54) are convex ones, and they can be efficiently solved by numerical optimization. We here introduce a well-known software for numerical convex optimization, CVXPY,[3] which is a Python library designed for numerical convex optimization. It provides a powerful and intuitive framework for constructing and solving convex optimization problems.

Let us consider the 80th-order polynomial $y = -t^{80} + t$ in (3.45). From this, we generate 11 data points as in (3.47), and we try to reconstruct the original polynomial from these data.

The Python program for ℓ^1 optimization is shown below.

```
import numpy as np
import matplotlib.pyplot as plt
import cvxpy as cp

# Polynomial coefficients
x_orig = np.zeros(80)
x_orig[0] = -1
x_orig[78] = 1

# Data
t = np.arange(0, 1.1, 0.1)
y = np.polyval(x_orig, t)

# Vandermonde matrix
N = len(t)
M = N - 1 # order of polynomial
Phi_v = np.vander(t)

## Interpolation polynomial
x_i = np.linalg.inv(Phi_v) @ y
print("Interpolating polynomial")
for coeff in x_i:
    print(f"{coeff:.4f}")
```

[3]https://www.cvxpy.org/

```
24
25 ## L1 optimization
26 # Vandermonde matrix
27 M_1 = 80 # order of polynomial
28 Phi_1 = np.vander(t, N=M_1+1)
29
30 # CVXPY for L1 optimization
31 x = cp.Variable(M_1+1)
32 objective = cp.Minimize(cp.norm(x, 1))
33 constraints = [Phi_1 @ x == y]
34 problem = cp.Problem(objective, constraints)
35 problem.solve()
36
37 # Extract the coefficients
38 x_lasso = x.value
39 print("\n LASSO")
40 for coeff in x.value:
41     print(f"{coeff:.4f}")
```

First, we import Python libraries `numpy`, `matplotlib.pyplot` and `cvxpy`, used in the program. Then, we define the coefficient vector of the 80th-order polynomial (lines 5–8). Next, by using a Numpy function `polyval` that returns the value of the polynomial from the coefficient vector, we make a data set (3.47) as in lines 10–12. With the data, we first find the 10th-order interpolating polynomial. For this, we compute the Vandermonde matrix (3.16) as in lines 14–17. From this, we compute the coefficients of the interpolating polynomial by using (3.20) as in line 20.

Figure 3.11 shows the curve of the obtained polynomial. In this case, the data are noiseless and there is no oscillation due to overfitting. However, we can see a large gap in the range from $t = 0.9$ to $t = 1$.

Finally, we compute the curve by ℓ^1 optimization (3.50). We assume that we know the polynomial order is at most 80. In this case, the Vandermonde matrix (3.16) becomes a fat matrix of size 11×81. This matrix can be obtained in lines 26–28. Define the coefficient vector $\boldsymbol{x}$ and the data vector $\boldsymbol{y}$ as in (3.17), then the condition for interpolation is described as

$$\boldsymbol{\Phi x} = \boldsymbol{y}. \tag{3.55}$$

We seek the sparsest solution in the set $\{\boldsymbol{x} \in \mathbb{R}^{81} : \boldsymbol{\Phi x} = \boldsymbol{y}\}$ by solving the ℓ^1 optimization problem in (3.50).

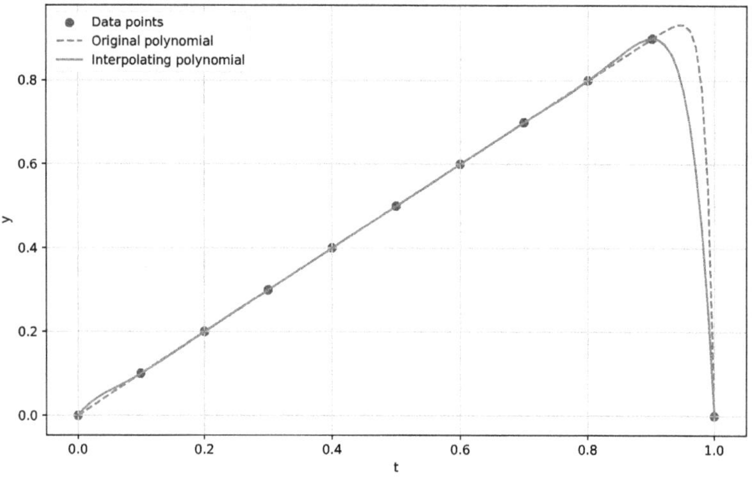

Figure 3.11: 10th-order interpolating polynomial (solid curve) and the original polynomial $y = -t^{80} + t$ (dashed curve)

To solve the ℓ^1 optimization problem, we use the CVXPY library. By using CVXPY, the optimization problem can be very easily coded as in lines 30–35. You should compare this program with (3.50). You can write a program very intuitively for an optimization problem. This is the strongest point of CVXPY. You can solve many convex optimization problems other than the ℓ^1 optimization in a similar way. We recommend for beginners to use CVXPY to solve convex optimization problems.[4]

Let us draw the curve with the coefficients obtained by the ℓ^1 optimization. Figure 3.12 shows the curve. We can see that the obtained curve is almost the same as the original curve $y = -t^{80} + t$. This is the power of ℓ^1 optimization.

3.3.2 Group testing

Let us consider the problem of group testing discussed in Section 2.4. Suppose there are $n = 1000$ individuals among which $s = 5$ individuals are infected. We assume the number of tests is $m = 100$. We choose the testing matrix $\Phi^{m \times n}$ as a random binary-valued matrix, that is, $\Phi \in \{0, 1\}^{m \times n}$. This means that the m groups are formed randomly. Then we obtain the measurement vector $\boldsymbol{y} = \Phi \boldsymbol{x}$, where $\boldsymbol{x}$ is a binary-valued vector that

[4]CVXPY is also available in MATLAB. See `https://cvxr.com/cvx/` for details.

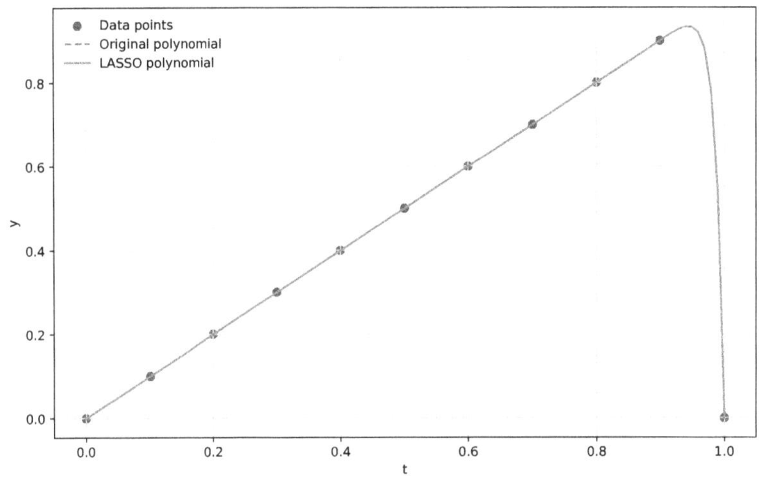

Figure 3.12: 80th-order polynomial by ℓ^1 optimization (solid curve) and the original polynomial $y = -t^{80} + t$ (dashed curve)

indicates the presence of infection in each individual, as defined in equations (2.39) and (2.40). Then the problem of group testing is formulated as the following ℓ^0 optimization:

$$\underset{\boldsymbol{x}\in\mathbb{R}^n}{\text{minimize}} \|\boldsymbol{x}\|_0 \quad \text{subject to} \quad \boldsymbol{y} = \Phi\boldsymbol{x}. \tag{3.56}$$

As discussed in Example 2.3, it takes a tremendous amount of time to solve this by the naive exhaustive search algorithm. Instead, we solve this by the ℓ^1 optimization:

$$\underset{\boldsymbol{x}\in\mathbb{R}^n}{\text{minimize}} \|\boldsymbol{x}\|_1 \quad \text{subject to} \quad \boldsymbol{y} = \Phi\boldsymbol{x}. \tag{3.57}$$

We numerically solve this optimization problem using CVXPY. The Python program is given at the end of this section. Figure 3.13 shows the results. The left figure shows the original binary vector of size $n = 1000$ including $s = 5$ ones. Then, by the ℓ^1 optimization, we obtain the reconstructed vector shown in the right figure. They are almost identical. The indices of the ones in the original vector are $\{37, 72, 235, 767, 908\}$, and the indices predicted from the reconstructed vector are exactly the same. The computational time is about 3 seconds (please check this by yourself!), and this is much faster than the exhaustive search.

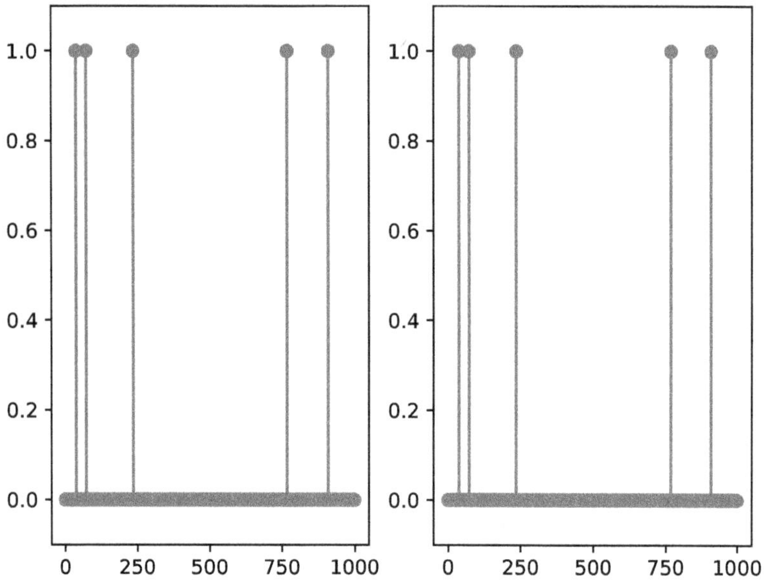

Figure 3.13: The original vector (left) and the reconstructed vector by ℓ^1 optimization (right)

```
import cvxpy as cp
import numpy as np
import matplotlib.pyplot as plt

## Parameter settings
# Vector size (number of individuals)
n = 1000
# Number of positives
s = 5
# Random seed
np.random.seed(1)
# Original vector (n-dimensional, k-sparse)
x_orig = np.zeros(n)
S = np.random.randint(n,size=s)
x_orig[S] = 1
# Number of tests
```

```
m = 100
# Testing matrix
A = np.random.randint(2,size=(m, n))
# Result vector
b = A @ x_orig

## Optimization by CVXPY
# Optimization variable
x = cp.Variable(n)
# Cost function (L1 norm)
cost = cp.norm1(x)
# Constraints (linear equations)
constraints = [A @ x == b]
# Optimization problem
prob = cp.Problem(cp.Minimize(cost), constraints)
# Solve by CVXPY
prob.solve()
# Print the result
print("status:", prob.status)
print("optimal value", prob.value)

## Results
fig = plt.figure()
ax1 = fig.add_subplot(1, 2, 1)
ax1.stem(x_orig)
plt.ylim(-0.1,1.1)
ax2 = fig.add_subplot(1, 2, 2)
ax2.stem(x.value)
plt.ylim(-0.1,1.1)
plt.show()

## Indeces
print(np.nonzero(x_orig))
x_est=np.round(x.value)
print(x_est.nonzero())
```

3.4 Further Readings

For the theory of regularization, I recommend reading standard textbooks [10], [51] of machine learning. The least squares method is deeply related to the projection and the generalized inverse, for which you can choose a textbook of [50]. The mathematical theory and generalization of LASSO can be found in [16], [47], [51], [52]. For the equivalence theorems between ℓ^0 and ℓ^1 optimizations, refer to textbooks [43], [45], [149].

Chapter 4

Algorithms for Convex Optimization

In the previous chapter, we have seen that convex optimization such as ℓ^1 optimization is efficiently solved by using CVXPY on Python. Such a tool is actually very useful for small or middle-scale problems. However, if you treat a very large-scale problem like image processing, CVXPY might be insufficient. Moreover, if you want to apply the ℓ^1 optimization to control systems, you should compute the optimal solution in real-time (e.g., in a few msec) with a cheap device on which Python and CVXPY cannot be installed. In such cases, you should instead write an efficient algorithm by yourself for your specific problem. This means that you should look into the *black box* of the toolbox. For this purpose, we review the basics of convex optimization, and introduce efficient algorithms for problems in compressed sensing.

Key ideas of Chapter 4

- In convex optimization, a local minimum is a global minimum.
- ℓ^1 optimization problems we study in this book are convex optimization.
- Proximal operators are used to derive fast algorithms for convex optimization with non-differentiable ℓ^1 norm and constraints.

4.1 Basics of Convex Optimization

We here review important facts in convex optimization. Let us begin with the definition of a convex set.

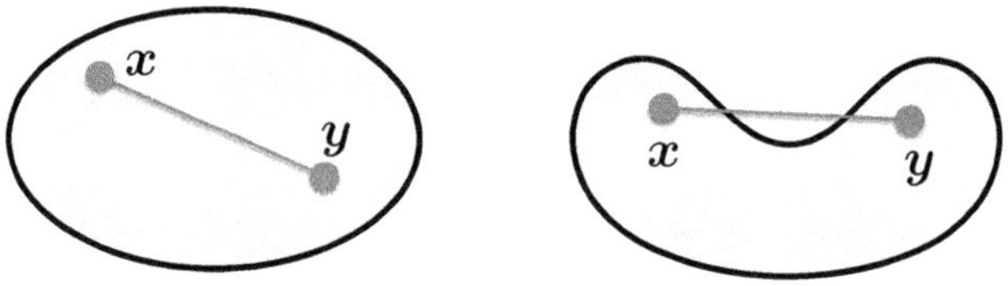

Figure 4.1: Convex set (left) and non-convex set (right)

Definition 4.1 (convex set). Let $\mathcal{C}$ be a subset of $\mathbb{R}^n$. $\mathcal{C}$ is said to be a *convex set* if the following inclusion

$$t\boldsymbol{x} + (1-t)\boldsymbol{y} \in \mathcal{C} \tag{4.1}$$

holds for any vectors $\boldsymbol{x}, \boldsymbol{y} \in \mathcal{C}$ and for any real number $t \in [0, 1]$.

Figure 4.1 illustrates a convex set and a *non-convex set* (i.e., a set that is not convex). In convex set $\mathcal{C}$, the line segment between any two points $\boldsymbol{x}$ and $\boldsymbol{y}$ in $\mathcal{C}$ lies completely in $\mathcal{C}$. On the other hand, in a non-convex set, there exists a line segment that partially lies outside of the set.

Exercise 4.1. Suppose that $\mathcal{C}$ and $\mathcal{D}$ are convex and $\mathcal{C} \cap \mathcal{D} \neq \emptyset$. Show that $\mathcal{C} \cap \mathcal{D}$ is convex.

In convex optimization, we often handle a function $f : \mathbb{R}^n \to \mathbb{R} \cup \{\infty\}$, which takes values on *extended real numbers* $\mathbb{R} \cup \{\infty\}$. The following function is an example:

$$f(\boldsymbol{x}) = \begin{cases} 0, & \text{if } \|\boldsymbol{x}\|_2 \leq 1, \\ \infty, & \text{if } \|\boldsymbol{x}\|_2 > 1. \end{cases} \tag{4.2}$$

This function is called an *indicator function*, which will be explained in Section 4.2.4.

The *effective domain* of a function f is defined by

$$\mathrm{dom}(f) \triangleq \{\boldsymbol{x} \in \mathbb{R}^n : f(\boldsymbol{x}) < \infty\}. \tag{4.3}$$

That is, the effective domain of a function $f : \mathbb{R}^n \to \mathbb{R} \cup \{\infty\}$ is a set in $\mathbb{R}^n$ on which f takes finite real values. For example, the effective domain of the indicator function (4.2) is given by

$$\mathrm{dom}(f) = \{\boldsymbol{x} \in \mathbb{R}^n : \|\boldsymbol{x}\|_2 \leq 1\}. \tag{4.4}$$

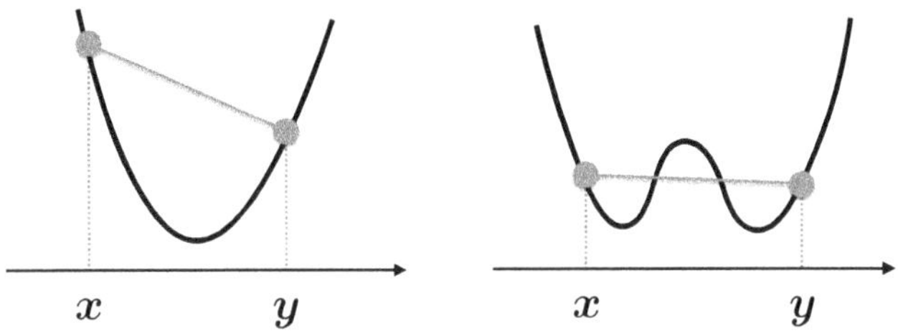

Figure 4.2: convex function (left) and non-convex function (right)

A function $f : \mathbb{R}^n \to \mathbb{R} \cup \{\infty\}$ is said to be *proper* if its effective domain is non-empty, that is, there exists at least one $\boldsymbol{x} \in \mathbb{R}^n$ such that $f(\boldsymbol{x}) < \infty$.

Now, let us define a convex function.

Definition 4.2 (convex function). Let $f : \mathbb{R}^n \to \mathbb{R} \cup \{\infty\}$ be a proper function. The function f is said to be a *convex function* if the following inequality

$$f(t\boldsymbol{x} + (1-t)\boldsymbol{y}) \leq tf(\boldsymbol{x}) + (1-t)f(\boldsymbol{y}) \tag{4.5}$$

holds for any vectors $\boldsymbol{x}, \boldsymbol{y} \in \mathrm{dom}(f)$ and for any real number $t \in [0, 1]$.

Figure 4.2 illustrates a convex function and a *non-convex function*, a function that is not convex. By definition, if f is convex, the line segment between any two points $(\boldsymbol{x}, f(\boldsymbol{x}))$ and $(\boldsymbol{y}, f(\boldsymbol{y}))$, where $\boldsymbol{x}, \boldsymbol{y} \in \mathrm{dom}(f)$, lies above or on the graph of f. On the other hand, if f is non-convex, there exists a line segment that partially lies below the graph.

Exercise 4.2. Suppose that f and g are convex functions and $\mathrm{dom}(f) \cap \mathrm{dom}(g) \neq \emptyset$. Show that $f + g$ is convex.

One more important property of a function is *closedness*. A function $f : \mathbb{R}^n \to \mathbb{R} \cup \{\infty\}$ is said to be *closed* if the *sublevel set* (or *lower level set*) $\{\boldsymbol{x} \in \mathrm{dom}(f) : f(\boldsymbol{x}) \leq c\}$ is a closed set for any $c \in \mathbb{R}$. The closedness of a function is also understood by its *epigraph*. The epigraph $\mathrm{epi}(f)$ of function f is defined by

$$\mathrm{epi}(f) \triangleq \{(\boldsymbol{x}, t) \in \mathbb{R}^n \times \mathbb{R} : \boldsymbol{x} \in \mathrm{dom}(f), f(\boldsymbol{x}) \leq t\}. \tag{4.6}$$

Figure 4.3 illustrates the epigraph of a function f. The epigraph of f is the region above the graph on its effective domain. It is easily shown that a function f is closed if and only if its epigraph is closed.

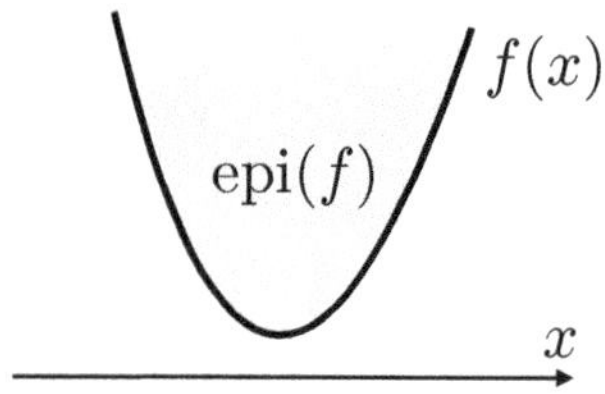

Figure 4.3: Epigraph epi(f) of function f

Table 4.1: Function and its epigraph

function f	epigraph epi(f)
convex	convex set
closed	closed set
proper	non-empty set

We can also perceive other properties of a function in terms of its epigraph. A function is convex if and only if its epigraph is convex. A function is proper if and only if its epigraph is non-empty. We summarize these facts in Table 4.1.

Now, we formulate a convex optimization problem in a general form.

Problem 4.1 (Convex optimization problem). Let $f : \mathbb{R}^n \to \mathbb{R} \cup \{\infty\}$ be a proper, closed, and convex function, and $\mathcal{C} \subset \mathbb{R}^n$ be a non-empty, closed, and convex set. Then, a *convex optimization problem* is a problem to find a vector $\boldsymbol{x}^* \in \mathbb{R}^n$ that minimizes the function $f(\boldsymbol{x})$ over the set $\mathcal{C} \subset \mathbb{R}^n$.

For the convex optimization, we use the following terminology in this book:

- The function $f(\boldsymbol{x})$ is called a *cost function* or an *objective function.*
- The set $\mathcal{C}$ is called a *constraint set* or a *feasible set.*
- The entries of $\mathcal{C}$ are called *feasible solutions.*
- The inclusion $\boldsymbol{x} \in \mathcal{C}$ is called a *constraint.*

The above optimization problem is often described as follows:

$$\underset{\boldsymbol{x}\in\mathbb{R}^n}{\text{minimize}} \;\; f(\boldsymbol{x}) \;\; \text{subject to} \;\; \boldsymbol{x} \in \mathcal{C}. \tag{4.7}$$

In this expression, the optimization variable $\boldsymbol{x} \in \mathbb{R}^n$ to be minimized is placed under "minimize", next to which the cost function $f(\boldsymbol{x})$ is placed.

$$\underset{\boldsymbol{x}\in\mathbb{R}^n}{\text{minimize}}\ \underbrace{f(\boldsymbol{x})}_{\text{cost function}}\ \text{subject to}\ \underbrace{\boldsymbol{x}\in C}_{\text{constraint}}$$

$$\min_{\boldsymbol{x}\in C} f(\boldsymbol{x}) \quad \text{minimum value}$$

$$\underset{\boldsymbol{x}\in C}{\arg\min}\ f(\boldsymbol{x}) \quad \text{minimizer (set)}$$

Figure 4.4: Notation for optimization problem

The term "subject to" is sometimes abbreviated as "s.t.", followed by the constraint $\boldsymbol{x} \in \mathcal{C}$. The term "minimize" in (4.7) is often abbreviated as "min" and simply described as

$$\min_{\boldsymbol{x}\in\mathbb{R}^n} f(\boldsymbol{x}) \quad \text{s.t.} \quad \boldsymbol{x} \in \mathcal{C}. \tag{4.8}$$

Also, we often write the constraint under "minimize" as

$$\min_{\boldsymbol{x}\in\mathcal{C}} f(\boldsymbol{x}). \tag{4.9}$$

Note that (4.9) sometimes means the *minimum value* of the optimization problem (4.7), instead of an optimization problem. The set of minimizers (solutions) to the optimization problem (4.7) is denoted using "arg" (abbreviation of argument) as

$$\arg\min_{\boldsymbol{x}\in\mathcal{C}} f(\boldsymbol{x}) \triangleq \{\boldsymbol{x}^* \in \mathcal{C} : f(\boldsymbol{x}^*) \le f(\boldsymbol{x}), \quad \forall \boldsymbol{x} \in \mathcal{C} \cap \operatorname{dom}(f)\}. \tag{4.10}$$

Also, we often use the following expression

$$\boldsymbol{x}^* = \arg\min_{\boldsymbol{x}\in\mathcal{C}} f(\boldsymbol{x}). \tag{4.11}$$

In this expression, "argmin" returns a minimizer, instead of the set of minimizers. If the minimizer of the optimization problem (4.7) is unique, then this expression may not cause any confusion. If not unique, (4.11) means that $\boldsymbol{x}^*$ is a minimizer arbitrarily taken from the set of minimizers.

We summarize the definitions in Figure 4.4.

Then we define a local minimizer and a global minimizer of the optimization problem (4.7). If there exists an open set $\mathcal{B} \subset \mathbb{R}^n$ that contains a feasible solution $\bar{\boldsymbol{x}} \in \mathcal{C}$ such that

$$f(\boldsymbol{x}) \ge f(\bar{\boldsymbol{x}}), \quad \forall \boldsymbol{x} \in \mathcal{B} \cap \mathcal{C}, \tag{4.12}$$

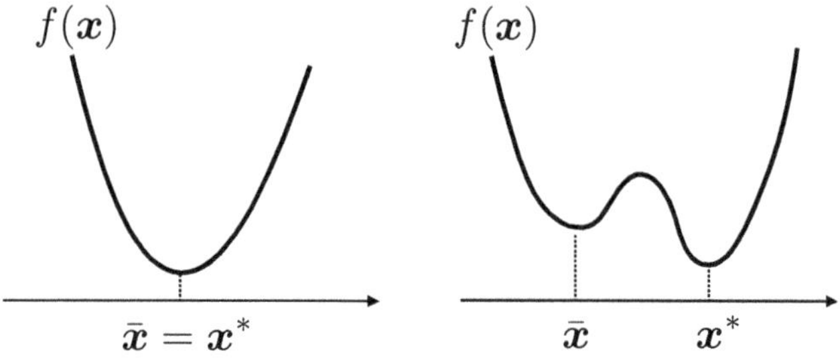

Figure 4.5: Local minimizer $\bar{\boldsymbol{x}}$ and global minimizer $\boldsymbol{x}^*$ with convex function (left) and non-convex function (right)

then $\bar{\boldsymbol{x}}$ is called a *local minimizer* of the optimization problem (4.7). If a feasible solution $\boldsymbol{x}^* \in C$ satisfies

$$f(\boldsymbol{x}) \geq f(\boldsymbol{x}^*), \quad \forall \boldsymbol{x} \in C, \tag{4.13}$$

then $\boldsymbol{x}^*$ is called a *global minimizer* of the optimization problem (4.7).

One of the most important properties of convex optimization is that a local minimizer is (if it exists) a global minimizer. Figure 4.5 illustrates this fact of convex optimization. In this figure, for a convex function, the local minimizer $\bar{\boldsymbol{x}}$ is also the global minimizer $\boldsymbol{x}^*$. On the other hand, for a non-convex function, they may not coincide. In fact, the following theorem holds [14, Section 4.2.2].

Theorem 4.1. For a convex optimization problem (4.7), any local minimizer is (if it exists) a global minimizer, and the set of global minimizers is a convex set.

By this theorem, an algorithm that outputs a local minimizer of a convex optimization problem is automatically an algorithm for a global minimizer. For example, a convex optimization problem with a differentiable and convex function $f(\boldsymbol{x})$ and $C = \mathbb{R}^n$ (unconstrained problem), a point $\bar{\boldsymbol{x}}$ such that $\nabla f(\bar{\boldsymbol{x}}) = \mathbf{0}$, where ∇f is the *gradient* of f, is a local minimizer, and this is also a global minimizer. Therefore, for an unconstrained convex optimization with a differentiable cost function, an algorithm searching for a point satisfying $\nabla f(\boldsymbol{x}) = \mathbf{0}$ is an algorithm for a global minimizer. Theorem 4.1 is very important to derive an efficient algorithm for convex optimization.

Exercise 4.3. Find a convex function that has no local minimizer.

Exercise 4.4. Find a convex function that has infinitely many local minimizers.

Next, we consider the uniqueness of the minimizer. For this, we define strictly and strongly convex functions.

Definition 4.3. Let $f : \mathbb{R}^n \to \mathbb{R} \cup \{\infty\}$ be a proper function. The function f is said to be a *strictly convex function* if for any $\boldsymbol{x}, \boldsymbol{y} \in \mathrm{dom}(f) \subset \mathbb{R}^n$ with $\boldsymbol{x} \neq \boldsymbol{y}$ and any $t \in (0, 1)$,

$$f(t\boldsymbol{x} + (1-t)\boldsymbol{y}) < tf(\boldsymbol{x}) + (1-t)f(\boldsymbol{y}) \tag{4.14}$$

holds. Moreover, the function f is said to be a *strongly convex function* if there exists $\beta > 0$ such that for any $\boldsymbol{x}, \boldsymbol{y} \in \mathrm{dom}(f) \subset \mathbb{R}^n$ and any $t \in [0, 1]$,

$$f(t\boldsymbol{x} + (1-t)\boldsymbol{y}) \leq tf(\boldsymbol{x}) + (1-t)f(\boldsymbol{y}) - t(1-t)\frac{\beta}{2}\|\boldsymbol{x} - \boldsymbol{y}\|_2^2 \tag{4.15}$$

holds. The constant β is called a *modulus*.

The following lemma is an important property of strongly convex functions.

Lemma 4.1. A function $f : \mathbb{R}^n \to \mathbb{R} \cup \{\infty\}$ is strongly convex with modulus $\beta > 0$ if and only if

$$f - \frac{\beta}{2}\|\cdot\|_2^2 \tag{4.16}$$

is convex.

Weierstrass extreme value theorem is also important in convex optimization.

Theorem 4.2 (Weierstrass extreme value theorem)**.** Every continuous function on a compact set attains its extreme values on that set.

Note that a subset in $\mathbb{R}^n$ is compact if and only if it is closed and bounded.

The following theorem shows the existence and uniqueness of the minimizer of a strongly convex function.

Theorem 4.3. Assume $f : \mathbb{R}^n \to \mathbb{R} \cup \{\infty\}$ is a proper, closed, and strongly convex function with modulus $\beta > 0$. Then f has the unique minimizer $\boldsymbol{x}^* \in \mathrm{dom}(f)$. That is, for any $\boldsymbol{x} \in \mathrm{dom}(f)$ such that $\boldsymbol{x} \neq \boldsymbol{x}^*$,

$$f(\boldsymbol{x}) > f(\boldsymbol{x}^*) \tag{4.17}$$

holds. Moreover, for any $\boldsymbol{x} \in \mathrm{dom}(f)$, we have

$$f(\boldsymbol{x}) \geq f(\boldsymbol{x}^*) + \frac{\beta}{2}\|\boldsymbol{x} - \boldsymbol{x}^*\|_2^2. \tag{4.18}$$

This theorem is used to define the proximal operator discussed in the next section.

Exercise 4.5. Prove Theorem 4.3.

4.2 Proximal Operators

We here introduce a powerful tool called the proximal operator for deriving efficient algorithms to solve convex optimization problems, particularly those involving non-differentiable cost functions.

4.2.1 Definition

The proximal operator of a function is defined as follows:

Definition 4.4 (proximal operator). For a proper, closed, and convex function $f : \mathbb{R}^n \to \mathbb{R} \cup \{\infty\}$, and a real number $\gamma > 0$, the *proximal operator* $\mathrm{prox}_{\gamma f}$ with parameter γ is defined by

$$\mathrm{prox}_{\gamma f}(\boldsymbol{v}) \triangleq \operatorname*{arg\,min}_{\boldsymbol{x} \in \mathrm{dom}(f)} \left\{ f(\boldsymbol{x}) + \frac{1}{2\gamma}\|\boldsymbol{x} - \boldsymbol{v}\|_2^2 \right\}. \tag{4.19}$$

First, we can easily show that the function

$$g(\boldsymbol{x}) \triangleq f(\boldsymbol{x}) + \frac{1}{2\gamma}\|\boldsymbol{x} - \boldsymbol{v}\|_2^2 \tag{4.20}$$

is a proper, closed, and strongly convex function with modulus $\beta = 1/\gamma$ (see Definition 4.3). Therefore, from Theorem 4.3, the proximal operator (4.19) is well-defined, that is, $\mathrm{prox}_{\gamma f}(\boldsymbol{v})$ uniquely exists for any $\boldsymbol{v} \in \mathbb{R}^n$.

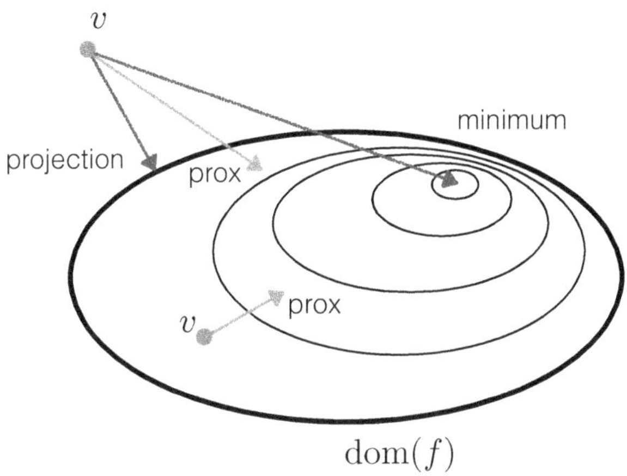

Figure 4.6: Illustration of proximal operator

Exercise 4.6. Assume that f is a proper, closed, and convex function, $\gamma > 0$, and $\boldsymbol{v} \in \mathbb{R}^n$. Prove that the function $g(\boldsymbol{x})$ in (4.20) is a proper, closed, and strongly convex function with modulus $\beta = 1/\gamma$.

From (4.19), if we take $\gamma \to \infty$, then the second term of (4.19) disappears and the proximal operator becomes

$$\text{prox}_{\infty f}(\boldsymbol{v}) = \underset{\boldsymbol{x} \in \text{dom}(f)}{\arg\min}\ f(\boldsymbol{x}) = \boldsymbol{x}^*, \tag{4.21}$$

where $\boldsymbol{x}^*$ is a minimizer of $f(\boldsymbol{x})$. On the other hand, taking $\gamma \to 0$ eliminates the first term of (4.19), and the proximal operator is reduced to

$$\text{prox}_{0f}(\boldsymbol{v}) = \underset{\boldsymbol{x} \in \text{dom}(f)}{\arg\min}\ \|\boldsymbol{x} - \boldsymbol{v}\|_2^2 = \Pi_{\mathcal{C}}(\boldsymbol{v}), \quad \mathcal{C} \triangleq \text{dom}(f), \tag{4.22}$$

where $\Pi_{\mathcal{C}}$ is the *projection operator* on the set $\mathcal{C}$. That is, $\Pi_{\mathcal{C}}$ returns the closest point in $\mathcal{C}$ measured by the ℓ^2 norm. Finally, if the parameter γ satisfies $0 < \gamma < \infty$, the proximal operator (4.19) is a mixture of the minimizer in (4.21) and the projection operator in (4.22).

Figure 4.6 illustrates the proximal operator. By definition, if a point $\boldsymbol{v}$ is outside the effective domain $\text{dom}(f)$, then $\text{prox}_{\gamma f}(\boldsymbol{v})$ moves into $\text{dom}(f)$. If a point $\boldsymbol{v}$ is in $\text{dom}(f)$, then $\text{prox}_{\gamma f}(\boldsymbol{v})$ moves in $\text{dom}(f)$, and approaches towards the minimizer $\boldsymbol{x}^*$ of $f(\boldsymbol{x})$, with a step size determined by the value of γ. Therefore, the effective domain $\text{dom}(f)$ is an *invariant set* under the proximal operator $\text{prox}_{\gamma f}$. Note that a set $\mathcal{C}$ is called an invariant set under an operator T if

$$\boldsymbol{x} \in \mathcal{C} \ \Rightarrow \ T(\boldsymbol{x}) \in \mathcal{C} \tag{4.23}$$

holds.

Exercise 4.7. Prove that the effective domain $\mathrm{dom}(f)$ is an invariant set under the proximal operator $\mathrm{prox}_{\gamma f}$.

As illustrated in Figure 4.6, if $\boldsymbol{v} \in \mathrm{dom}(f)$, then the vector $\mathrm{prox}_{\gamma f}(\boldsymbol{v})$ approaches the minimizer $\boldsymbol{x}^*$ within the effective domain $\mathrm{dom}(f)$. That is, a proximal operator behaves similarly to the *negative gradient* of f (if f is differentiable) within the effective domain. The proximal operator can be applied to a broader class of functions, including those that are not differentiable.

4.2.2 Proximal algorithm

From the invariance property of (4.23), we can consider an iterative algorithm called the *proximal algorithm* that seeks a minimizer of convex function f:

Proximal algorithm

Initialization: give an initial vector $\boldsymbol{x}[0]$ and positive numbers $\gamma_0, \gamma_1, \ldots$

Iteration: for $k = 0, 1, 2, \ldots,$ do

$$\boldsymbol{x}[k+1] = \mathrm{prox}_{\gamma_k f}(\boldsymbol{x}[k]). \tag{4.24}$$

If you properly choose the parameter sequence $\{\gamma_k\}$, you can obtain one of the minimizers of f by the proximal algorithm. The convergence is shown in the following theorem [9, Proposition 5.1.3]:

Theorem 4.4 (convergence of proximal algorithm). Suppose that the parameter sequence $\{\gamma_k\}$ satisfies $\gamma_k > 0$ for all k and

$$\sum_{k=0}^{\infty} \gamma_k = \infty. \tag{4.25}$$

Then, the vector sequence $\{\boldsymbol{x}[k]\}$ generated by the proximal algorithm (4.24) converges to one of the minimizers of f for any initial vector $\boldsymbol{x}[0]$.

The theorem is based on the fact that a minimizer of $f(\boldsymbol{x})$ is also a *fixed point* of its proximal operator $\mathrm{prox}_{\gamma f}$. Note that a fixed point of $\mathrm{prox}_{\gamma f}$ is a point that satisfies

$$\boldsymbol{x} = \text{prox}_{\gamma f}(\boldsymbol{x}). \tag{4.26}$$

A fixed point is literally *fixed* under the operation by $\text{prox}_{\gamma f}$.

The proximal algorithm minimizes the *strongly convex* function

$$g_k(\boldsymbol{x}) \triangleq f(\boldsymbol{x}) + \frac{1}{2\gamma_k}\|\boldsymbol{x} - \boldsymbol{x}[k]\|_2^2 \tag{4.27}$$

at step k. In other words, the algorithm approximates a general convex function $f(\boldsymbol{x})$ by a strongly convex function at each step.

Also, it is often important to find a closed form of the proximal operator (4.19) for an efficient algorithm. A function for which the proximal operator is obtained in a closed form is sometimes called *proximable.* Let us see some proximable functions in the following subsections.

4.2.3 Proximal operator for quadratic function

Let us consider the following *quadratic function*

$$f(\boldsymbol{x}) = \frac{1}{2}\boldsymbol{x}^\top \Phi \boldsymbol{x} - \boldsymbol{y}^\top \boldsymbol{x}, \tag{4.28}$$

where Φ is a *positive-definite* symmetric matrix. Note that a symmetric matrix Φ is said to be *positive definite* if the following inequality holds

$$\boldsymbol{x}^\top \Phi \boldsymbol{x} > 0, \tag{4.29}$$

for every nonzero vector $\boldsymbol{x} \in \mathbb{R}^n$. Let us compute the proximal operator of the quadratic function in (4.28). From the definition (4.19) of the proximal operator, we have

$$\text{prox}_{\gamma f}(\boldsymbol{v}) = \arg\min_{\boldsymbol{x}\in\mathbb{R}^n} \left\{ \frac{1}{2}\boldsymbol{x}^\top \Phi \boldsymbol{x} - \boldsymbol{y}^\top \boldsymbol{x} + \frac{1}{2\gamma}(\boldsymbol{x} - \boldsymbol{v})^\top (\boldsymbol{x} - \boldsymbol{v}) \right\}. \tag{4.30}$$

Since the function in (4.30) is differentiable, we can obtain the minimizer by setting the gradient to be zero. After some calculations, we have the proximal operator in a closed form:

$$\text{prox}_{\gamma f}(\boldsymbol{v}) = \left(\Phi + \frac{1}{\gamma}I\right)^{-1}\left(\boldsymbol{y} + \frac{1}{\gamma}\boldsymbol{v}\right). \tag{4.31}$$

Exercise 4.8. Prove that the equation (4.31) holds.

An important application of this proximal operator is numerical matrix inversion. The minimizer $\boldsymbol{x}^*$ of (4.28) is also the unique solution to the linear equation

$$\Phi \boldsymbol{x} = \boldsymbol{y}, \tag{4.32}$$

that is, $\boldsymbol{x}^* = \Phi^{-1}\boldsymbol{y}$. Note that Φ is invertible since Φ is positive definite. Then, let us assume that the *condition number* of Φ, the ratio of its maximum and minimum eigenvalues, is very large so that the numerical computation of the inverse is difficult. We call such a case *ill-conditioned.* For an ill-conditioned case, the proximal algorithm (4.24) is used to safely compute the inverse. From (4.31), the proximal algorithm to obtain the minimizer of (4.28), which is also the solution to (4.32), is given as follows:

Proximal algorithm for $\Phi^{-1}\boldsymbol{y}$

Initialization: give an initial vector $\boldsymbol{x}[0]$ and a positive number $\gamma > 0$.
Iteration: for $k = 0, 1, 2, \ldots,$ do

$$\boldsymbol{x}[k+1] = \left(\Phi + \frac{1}{\gamma}I\right)^{-1}\left(\boldsymbol{y} + \frac{1}{\gamma}\boldsymbol{x}[k]\right). \tag{4.33}$$

If the positive number γ is sufficiently small, then the condition number of matrix $\Phi + (1/\gamma)I$ is relatively small, and the inversion can be easily computed numerically.

Also, if γ is sufficiently small, we have

$$\left(\Phi + \frac{1}{\gamma}I\right)^{-1} = \gamma(I + \gamma\Phi)^{-1} \approx \gamma(I - \gamma\Phi). \tag{4.34}$$

Then, the right-hand side of the proximal operator (4.31) becomes

$$\begin{aligned}\left(\Phi + \frac{1}{\gamma}I\right)^{-1}\left(\boldsymbol{y} + \frac{1}{\gamma}\boldsymbol{v}\right) &\approx \gamma(I - \gamma\Phi)\left(\boldsymbol{y} + \frac{1}{\gamma}\boldsymbol{v}\right)\\ &\approx \boldsymbol{v} - \gamma(\Phi\boldsymbol{v} - \boldsymbol{y})\\ &= \boldsymbol{v} - \gamma\nabla f(\boldsymbol{v}).\end{aligned} \tag{4.35}$$

That is, if γ is sufficiently small, the proximal algorithm (4.33) approximates the behavior of the *gradient descent algorithm*

$$\boldsymbol{x}[k+1] = \boldsymbol{x}[k] - \gamma\nabla f(\boldsymbol{x}[k]), \quad k = 0, 1, 2, \ldots, \tag{4.36}$$

to find the minimizer of the quadratic function (4.28).

4.2.4 Proximal operator for indicator function

The *indicator function* of a non-empty set $\mathcal{C} \subset \mathbb{R}^n$ is defined by

$$I_{\mathcal{C}}(\boldsymbol{x}) \triangleq \begin{cases} 0, & \text{if } \boldsymbol{x} \in \mathcal{C},\\ \infty, & \text{if } \boldsymbol{x} \notin \mathcal{C}.\end{cases} \tag{4.37}$$

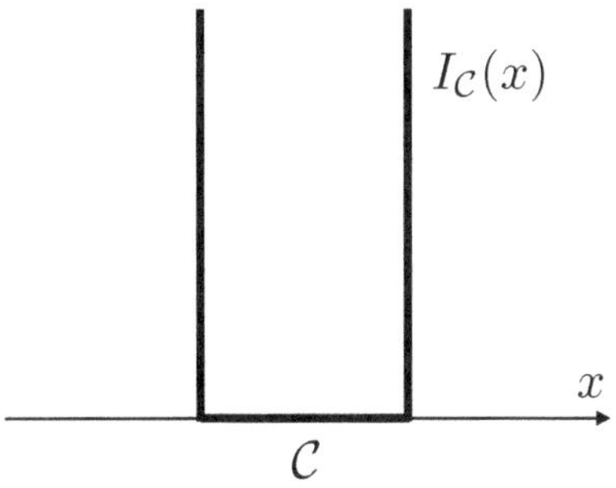

Figure 4.7: Indicator function $I_{\mathcal{C}}(x)$ on a closed interval $\mathcal{C} \in \mathbb{R}$

If the set $\mathcal{C}$ is non-empty, closed, and convex, then the indicator function $I_{\mathcal{C}}(\boldsymbol{x})$ is a proper, closed, and convex function (to check this, draw the epigraph). For example, the indicator function $I_{\mathcal{C}}(x)$ of a closed interval $\mathcal{C}$ on $\mathbb{R}$ is illustrated in Figure 4.7. You can see that if $\mathcal{C}$ is a non-empty closed interval, the epigraph is a non-empty, closed, and convex set.

Let us compute the proximal operator of the indicator function $I_{\mathcal{C}}$. From the definition (4.19), the proximal operator of $I_{\mathcal{C}}$ is given as

$$\begin{aligned} \operatorname{prox}_{\gamma I_{\mathcal{C}}}(\boldsymbol{v}) &= \operatorname*{arg\,min}_{\boldsymbol{x} \in \mathbb{R}^n} \left\{ I_{\mathcal{C}}(\boldsymbol{x}) + \frac{1}{2\gamma} \|\boldsymbol{x} - \boldsymbol{v}\|_2^2 \right\} \\ &= \operatorname*{arg\,min}_{\boldsymbol{x} \in \mathcal{C}} \|\boldsymbol{x} - \boldsymbol{v}\|_2^2 \\ &= \Pi_{\mathcal{C}}(\boldsymbol{v}). \end{aligned} \tag{4.38}$$

That is, the proximal operator of the indicator function $I_{\mathcal{C}}$ is the *projection operator* $\Pi_{\mathcal{C}}$ onto the set $\mathcal{C}$.

Exercise 4.9. Suppose that $\mathcal{C} \subset \mathbb{R}^n$ is a non-empty, closed, and convex set. Prove that $\Pi_{\mathcal{C}}(\boldsymbol{v})$ is uniquely determined for any $\boldsymbol{v} \in \mathbb{R}^n$.

4.2.5 Proximal operator for ℓ^1 norm

Let us compute the proximal operator (4.19) for the ℓ^1 norm:

$$f(\boldsymbol{x}) = \|\boldsymbol{x}\|_1 = \sum_{i=1}^{n} |x_i|, \tag{4.39}$$

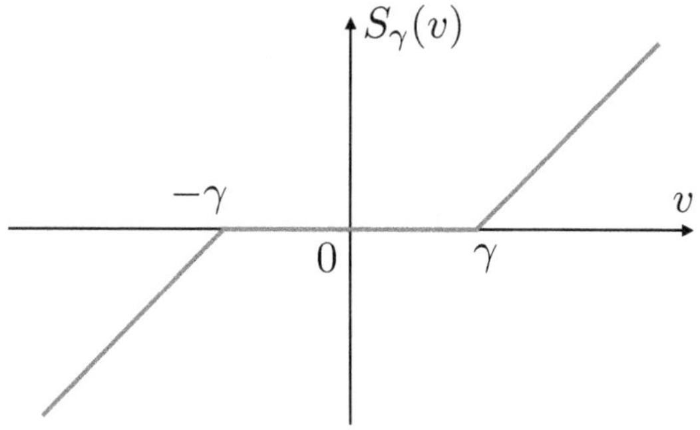

Figure 4.8: Soft-thresholding operator $S_\gamma(v)$

where x_i is the i-th element of $\boldsymbol{x} \in \mathbb{R}^n$. From the definition (4.19) of the proximal operator, we have

$$\begin{aligned} \operatorname{prox}_{\gamma\|\cdot\|_1}(\boldsymbol{v}) &= \operatorname*{arg\,min}_{\boldsymbol{x}\in\mathbb{R}^n} \left\{ \|\boldsymbol{x}\|_1 + \frac{1}{2\gamma}\|\boldsymbol{x}-\boldsymbol{v}\|_2^2 \right\} \\ &= \operatorname*{arg\,min}_{\boldsymbol{x}\in\mathbb{R}^n} \sum_{i=1}^n \left\{ |x_i| + \frac{1}{2\gamma}(x_i - v_i)^2 \right\}, \end{aligned} \tag{4.40}$$

where v_i is the i-th element of $\boldsymbol{v}$. This optimization can be reduced to element-wise optimization, that is,

$$\min_{\boldsymbol{x}\in\mathbb{R}^n} \sum_{i=1}^n \left\{ |x_i| + \frac{1}{2\gamma}(x_i - v_i)^2 \right\} = \sum_{i=1}^n \min_{x_i\in\mathbb{R}} \left\{ |x_i| + \frac{1}{2\gamma}(x_i - v_i)^2 \right\}. \tag{4.41}$$

Therefore, we just solve the following scalar minimization problem:

$$\operatorname*{minimize}_{x\in\mathbb{R}} \ |x| + \frac{1}{2\gamma}(x-v)^2. \tag{4.42}$$

The minimizer $x^* \in \mathbb{R}$ can be easily calculated, which is given by

$$x^* = S_\gamma(v) \triangleq \begin{cases} v - \gamma, & \text{if } \ v \geq \gamma, \\ 0, & \text{if } \ -\gamma < v < \gamma, \\ v + \gamma, & \text{if } \ v \leq -\gamma. \end{cases} \tag{4.43}$$

The function $S_\gamma(v)$ in (4.43) is called the *soft-thresholding operator*. Figure 4.8 shows the graph of this operator.

Exercise 4.10. Show that the minimizer x^* of the function

$$f(x) \triangleq |x| + \frac{1}{2\gamma}(x-v)^2 \tag{4.44}$$

is given by (4.43). (Hint: divide the domain of $f(x)$ into two intervals: $x \geq 0$ and $x < 0$. Then, consider the three cases for v: $v \geq \gamma$, $-\gamma < v < \gamma$, and $v \leq -\gamma$.)

By using the scalar-valued soft-thresholding operator, the proximal operator of the ℓ^1 norm is given by

$$\left[\mathrm{prox}_{\gamma f}(\boldsymbol{v})\right]_i = S_\gamma(v_i), \tag{4.45}$$

where $[\ \ ]_i$ denotes the i-th element of the vector in the square bracket. For a simple expression, we extend the definition of the scalar-valued soft-thresholding operator (4.43) to vectors. For a vector $\boldsymbol{v} \in \mathbb{R}^n$, we define the vector-valued soft-thresholding operator $S_\gamma(\boldsymbol{v})$ by

$$[S_\gamma(\boldsymbol{v})]_i \triangleq S_\gamma(v_i), \tag{4.46}$$

where $[S_\gamma(\boldsymbol{v})]_i$ is the i-th element of $S_\gamma(\boldsymbol{v})$. With this notation, the proximal operator of the ℓ^1 norm (4.45) is simply rewritten as

$$\mathrm{prox}_{\gamma\|\cdot\|_1}(\boldsymbol{v}) = S_\gamma(\boldsymbol{v}). \tag{4.47}$$

Exercise 4.11. Let $Q \in \mathbb{R}^{n\times n}$ be an orthogonal matrix. Prove that the minimizer $\boldsymbol{x}^* \in \mathbb{R}^n$ of the following function

$$f(\boldsymbol{x}) \triangleq \frac{1}{2}\|Q\boldsymbol{x} - \boldsymbol{y}\|_2^2 + \lambda\|\boldsymbol{x}\|_1 \tag{4.48}$$

is given by

$$\boldsymbol{x}^* = S_\lambda(Q^\top \boldsymbol{y}). \tag{4.49}$$

Note that Q is orthogonal if and only if

$$QQ^\top = Q^\top Q = I. \tag{4.50}$$

In summary, the proximal operator of the ℓ^1 norm is the soft-thresholding operator $S_\gamma(\boldsymbol{v})$. If the absolute value of an element v_i in $\boldsymbol{v}$ is less than γ, then the element is set to be zero by the proximal operator. This is an important property to understand why ℓ^1 optimization gives a sparse solution.

The word 'soft' means that the operator is continuous (see Figure 4.8). We can also define the *hard*-thresholding operator by

$$H_\lambda(v) \triangleq \begin{cases} v, & \text{if } |v| \geq \lambda, \\ 0, & \text{if } |v| < \lambda. \end{cases} \tag{4.51}$$

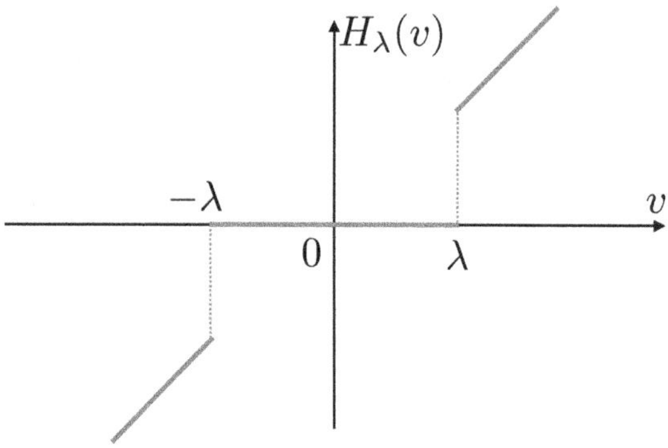

Figure 4.9: Hard-thresholding operator $H_\lambda(v)$

Figure 4.9 shows the graph of this operator. We can see from this figure, the hard-thresholding operator is discontinuous. An interesting fact is that the hard-thresholding operator is the proximal operator of the ℓ^0 norm with $\lambda = \sqrt{2\gamma}$. Strictly speaking, this is dubious since the proximal operator is defined for proper, closed, and convex functions (see Definition 4.4), but the ℓ^0 norm is not convex. However, the hard-thresholding operator is very useful to derive efficient algorithms for ℓ^0-norm optimization. See Chapter 5 for details.

Exercise 4.12. Compute the proximal operator (4.19) of the ℓ^0 norm, and show that it is the hard-thresholding operator (4.51).

4.3 Proximal Splitting Methods for ℓ^1 Optimization

In this section, we derive an efficient algorithm based on proximal splitting to solve the ℓ^1 optimization:

$$\underset{\boldsymbol{x}\in\mathbb{R}^n}{\text{minimize}} \ \|\boldsymbol{x}\|_1 \quad \text{subject to} \quad \Phi\boldsymbol{x} = \boldsymbol{y}, \tag{4.52}$$

where $\Phi \in \mathbb{R}^{m\times n}$ and $\boldsymbol{y} \in \mathbb{R}^m$ are given. We assume that $m < n$ and Φ has full row rank, that is, $\text{rank}(\Phi) = m$. The cost function is the ℓ^1 norm, which is obviously a proper, closed, and convex function. Then, let us consider the constraint. Let $\mathcal{C}$ denote the set of vectors $\boldsymbol{x} \in \mathbb{R}^n$ satisfying the constraint $\Phi\boldsymbol{x} = \boldsymbol{y}$. That is,

$$\mathcal{C} \triangleq \{\boldsymbol{x} \in \mathbb{R}^n : \Phi\boldsymbol{x} = \boldsymbol{y}\}. \tag{4.53}$$

It is easy to prove that this set is a non-empty, closed, and convex set in $\mathbb{R}^n$ if $\text{rank}(\Phi) = m$.

Exercise 4.13. Show the cost function $\|\boldsymbol{x}\|_1$ in (4.52) is a proper, closed, and convex function. Also, show the set $\mathcal{C}$ defined in (4.53) is a non-empty, closed, and convex set in $\mathbb{R}^n$.

Then, consider the indicator function $I_{\mathcal{C}}(\boldsymbol{x})$ for $\mathcal{C}$:

$$I_{\mathcal{C}}(\boldsymbol{x}) = \begin{cases} 0, & \text{if } \Phi\boldsymbol{x} = \boldsymbol{y}, \\ \infty, & \text{if } \Phi\boldsymbol{x} \neq \boldsymbol{y}. \end{cases} \tag{4.54}$$

By using this, the optimization problem in (4.52) is equivalently rewritten as

$$\underset{\boldsymbol{x}\in\mathbb{R}^n}{\text{minimize}} \ \|\boldsymbol{x}\|_1 + I_{\mathcal{C}}(\boldsymbol{x}). \tag{4.55}$$

Note that the functions $\|\boldsymbol{x}\|_1$ and $I_{\mathcal{C}}(\boldsymbol{x})$ are both proper, closed, and convex functions, and hence the sum of them, $\|\boldsymbol{x}\|_1 + I_{\mathcal{C}}(\boldsymbol{x})$, is also proper, closed, and convex.

Exercise 4.14. Suppose that two functions, f and g, are proper, closed, and convex. Suppose also that $\text{dom}(f) \cap \text{dom}(g) \neq \emptyset$. Prove that $f + g$ is also a proper, closed, and convex function.

Note that for the optimization problem (4.55), we cannot obtain the proximal operator of the cost function

$$f(\boldsymbol{x}) \triangleq \|\boldsymbol{x}\|_1 + I_{\mathcal{C}}(\boldsymbol{x}), \tag{4.56}$$

in a closed form. In other words, $f(\boldsymbol{x})$ is not proximable. That is, we cannot directly apply the proximal algorithm (4.25) to this problem. However, the proximal operators of the two functions

$$f_1(\boldsymbol{x}) \triangleq \|\boldsymbol{x}\|_1, \quad f_2(\boldsymbol{x}) \triangleq I_{\mathcal{C}}(\boldsymbol{x}) \tag{4.57}$$

can be obtained as the soft-thresholding operator in (4.47) and the projection operator onto $\mathcal{C}$ defined in (4.38), respectively. The idea is to *split* the cost function as $f = f_1 + f_2$, and write an algorithm using the proximal operators of f_1 and f_2 separately. Algorithms designed by this idea are called *proximal splitting algorithms.* For the problem (4.55), we introduce two proximal splitting algorithms in the following subsections.

4.3.1 Douglas-Rachford splitting algorithm

Let us consider the following optimization problem in a general form:

$$\underset{\boldsymbol{x}\in\mathbb{R}^n}{\text{minimize}} \ f_1(\boldsymbol{x}) + f_2(\boldsymbol{x}), \tag{4.58}$$

where f_1 and f_2 are proper, closed, and convex functions. The *Douglas-Rachford splitting algorithm* for (4.58) is given as follows:

Douglas-Rachford splitting algorithm for (4.58)

Initialization: give an initial vector $\boldsymbol{z}[0]$ and a parameter $\gamma > 0$.
Iteration: for $k = 0, 1, 2, \dots$ do

$$\begin{aligned}\boldsymbol{x}[k+1] &= \operatorname{prox}_{\gamma f_1}(\boldsymbol{z}[k]),\\ \boldsymbol{z}[k+1] &= \boldsymbol{z}[k] + \operatorname{prox}_{\gamma f_2}(2\boldsymbol{x}[k+1] - \boldsymbol{z}[k]) - \boldsymbol{x}[k+1].\end{aligned} \tag{4.59}$$

From the algorithm, we can derive an algorithm for our unconstrained problem (4.55). In our case, $f_1(\boldsymbol{x}) = \|\boldsymbol{x}\|_1$ and $f_2(\boldsymbol{x}) = I_{\mathcal{C}}(\boldsymbol{x})$, for which the proximal operators are given by

$$\operatorname{prox}_{\gamma f_1}(\boldsymbol{v}) = S_\gamma(\boldsymbol{v}), \quad \operatorname{prox}_{\gamma f_2}(\boldsymbol{v}) = \Pi_{\mathcal{C}}(\boldsymbol{v}). \tag{4.60}$$

Then the Douglas-Rachford splitting algorithm for the ℓ^1 optimization problem (4.52) is given as follows:

Douglas-Rachford splitting algorithm for (4.52)

Initialization: give an initial vector $\boldsymbol{z}[0]$ and a parameter $\gamma > 0$.
Iteration: for $k = 0, 1, 2, \dots$ do

$$\begin{aligned}\boldsymbol{x}[k+1] &= S_\gamma(\boldsymbol{z}[k]),\\ \boldsymbol{z}[k+1] &= \boldsymbol{z}[k] + \Pi_{\mathcal{C}}(2\boldsymbol{x}[k+1] - \boldsymbol{z}[k]) - \boldsymbol{x}[k+1].\end{aligned} \tag{4.61}$$

In this algorithm, the projection operator $\Pi_{\mathcal{C}}$ on the hyperplane $\mathcal{C}$ defined in (4.53) is given by

$$\Pi_{\mathcal{C}}(\boldsymbol{v}) = \boldsymbol{v} + \Phi^\top(\Phi\Phi^\top)^{-1}(\boldsymbol{y} - \Phi\boldsymbol{v}). \tag{4.62}$$

Note that $\Phi\Phi^\top$ is invertible since Φ has full row rank.

Exercise 4.15. Show that the projection operator $\Pi_{\mathcal{C}}$ for $\mathcal{C}$ defined in (4.53) is given by (4.62).

The ℓ^1 optimization problem (4.52) can be rewritten as a linear programming problem, which can be efficiently solved by the well-known interior-point method [54, Section 5.12.]. However, this method should solve a system of linear equations at each step of the iteration, which takes in general non-negligible computational time. On the other hand, the Douglas-Rachford algorithm in (4.61) only requires

- simple continuous mapping of the soft-thresholding function S_γ,
- and linear computation of matrix-vector multiplication and vector addition.

Thus, the Douglas-Rachford algorithm is efficient and easy to implement compared to standard interior-point algorithms.

To consider the convergence of Douglas-Rachford splitting algorithm, we define the *relative interior* $\mathrm{ri}(\mathcal{C})$ of a subset $\mathcal{C} \subset \mathbb{R}^n$ by

$$\mathrm{ri}(\mathcal{C}) \triangleq \{\boldsymbol{x} \in \mathbb{R}^n : \boldsymbol{x} \in \mathcal{C} \ \text{ and } \ \exists\, \epsilon > 0,\ \mathcal{N}_\epsilon(\boldsymbol{x}) \cap \mathrm{aff}(\mathcal{C}) \subset \mathcal{C}\}, \tag{4.63}$$

where $\mathcal{N}_\epsilon(\boldsymbol{x})$ is the ϵ-neighborhood of $\boldsymbol{x}$, that is,

$$\mathcal{N}_\epsilon \triangleq \{\boldsymbol{v} \in \mathbb{R}^n : \|\boldsymbol{v} - \boldsymbol{x}\|_2 < \epsilon\}, \tag{4.64}$$

and $\mathrm{aff}(\mathcal{C})$ is the affine hull of $\mathcal{C}$, that is, the set of all affine sets containing $\mathcal{C}$. Note that the relative interior is different from the *interior* of $\mathcal{C}$ that is defined by

$$\mathrm{int}(\mathcal{C}) \triangleq \{\boldsymbol{x} \in \mathbb{R}^n : \boldsymbol{x} \in \mathcal{C} \ \text{ and } \ \exists\, \epsilon > 0, \mathcal{N}_\epsilon(\boldsymbol{x}) \subset \mathcal{C}\}. \tag{4.65}$$

For example, let us consider the disc

$$\mathcal{C} = \{(x_1, x_2, 0) \in \mathbb{R}^3 : x_1^2 + x_2^2 \le 1\}, \tag{4.66}$$

on the x_1-x_2 plane in $\mathbb{R}^3$. Then, the interior of $\mathcal{C}$ is empty by definition (4.65), while the relative interior is

$$\mathrm{ri}(\mathcal{C}) = \{(x_1, x_2, 0) \in \mathbb{R}^3 : x_1^2 + x_2^2 < 1\}. \tag{4.67}$$

Now, we introduce the convergence theorem [29] for the Douglas-Rachford splitting algorithm.

Theorem 4.5. Suppose that f_1 and f_2 are proper, closed, and convex functions that satisfy

$$\mathrm{ri}(\mathrm{dom}(f_1)) \cap \mathrm{ri}(\mathrm{dom}(f_2)) \neq \emptyset. \tag{4.68}$$

Also, suppose that

$$f_1(\boldsymbol{x}) + f_2(\boldsymbol{x}) \to \infty \ \text{ as } \ \|\boldsymbol{x}\|_2 \to \infty. \tag{4.69}$$

Then each sequence $\{\boldsymbol{x}[k]\}_{k=0}^{\infty}$ generated by the Douglas-Rachford splitting algorithm converges to a solution to the optimization problem (4.58).

4.3.2 Dykstra-like splitting algorithm

Here we compute the proximal operator of $f = f_1 + f_2$ *numerically.* Let us consider the following optimization problem:

$$\underset{\boldsymbol{x}\in\mathbb{R}^n}{\text{minimize}}\ f_1(\boldsymbol{x}) + f_2(\boldsymbol{x}) + \frac{1}{2}\|\boldsymbol{x}-\boldsymbol{v}\|_2^2, \tag{4.70}$$

where f_1 and f_2 are proper, closed, and convex functions. The solution is given by $\boldsymbol{x}^* = \text{prox}_{\gamma(f_1+f_2)}(\boldsymbol{v})$ with $\gamma = 1$, but we assume $f_1 + f_2$ is not proximable. To solve this, we adopt the Douglas-Rachford splitting algorithm by splitting the cost function into f_1 and $f_2 + \frac{1}{2}\|\cdot - \boldsymbol{v}\|_2^2$. Then an algorithm called *Dykstra-like splitting algorithm* is obtained as follows:

Dykstra-like splitting algorithm for (4.70)

Initialization: set $\boldsymbol{x}[0] = \boldsymbol{v}$ and $\boldsymbol{p}[0] = \boldsymbol{q}[0] = \boldsymbol{0}$; give a parameter $\gamma > 0$.
Iteration: for $k = 0, 1, 2, \ldots$ do

$$\begin{aligned} \boldsymbol{z}[k] &= \text{prox}_{\gamma f_2}(\boldsymbol{x}[k] + \boldsymbol{p}[k]), \\ \boldsymbol{p}[k+1] &= \boldsymbol{x}[k] + \boldsymbol{p}[k] - \boldsymbol{z}[k], \\ \boldsymbol{x}[k+1] &= \text{prox}_{\gamma f_1}(\boldsymbol{z}[k] + \boldsymbol{q}[k]), \\ \boldsymbol{q}[k+1] &= \boldsymbol{z}[k] + \boldsymbol{q}[k] - \boldsymbol{x}[k+1]. \end{aligned} \tag{4.71}$$

An important application of the Dykstra-like algorithm is to find the projection of a point onto the intersection of two convex sets $\mathcal{C}_1$ and $\mathcal{C}_2$, namely to find $\Pi_{\mathcal{C}_1\cap\mathcal{C}_2}(\boldsymbol{v})$. This is done by setting $f_1 = I_{\mathcal{C}_1}$ and $f_2 = I_{\mathcal{C}_2}$, indicator functions defined in (4.37), for the optimization problem (4.70). Since $\text{prox}_{\gamma I_{\mathcal{C}_1}} = \Pi_{\mathcal{C}_1}$ and $\text{prox}_{\gamma I_{\mathcal{C}_2}} = \Pi_{\mathcal{C}_2}$, the algorithm is given by

$$\begin{aligned} \boldsymbol{z}[k+1] &= \Pi_{\mathcal{C}_1}(\boldsymbol{x}[k] + \boldsymbol{p}[k]), \\ \boldsymbol{p}[k+1] &= \boldsymbol{x}[k] + \boldsymbol{p}[k] - \boldsymbol{z}[k], \\ \boldsymbol{x}[k+1] &= \Pi_{\mathcal{C}_2}(\boldsymbol{z}[k] + \boldsymbol{q}[k]), \\ \boldsymbol{q}[k+1] &= \boldsymbol{z}[k] + \boldsymbol{q}[k] - \boldsymbol{x}[k+1]. \end{aligned} \tag{4.72}$$

This is called the *Dykstra projection algorithm*, proposed by Dykstra [15].[1] The name "Dykstra-like splitting" is actually after this algorithm. The convergence theorem is given as follows [29].

[1]Note that Dykstra for this algorithm is different from *Dijkstra* who found a famous algorithm for a shortest path in a network.

Theorem 4.6. Suppose that f_1 and f_2 are proper, closed, and convex functions that satisfy

$$\mathrm{dom}(f_1) \cap \mathrm{dom}(f_2) \neq \emptyset. \tag{4.73}$$

Then each sequence $\{\boldsymbol{x}[k]\}_{k=0}^{\infty}$ generated by the Dykstra-like splitting algorithm (4.71) converges to a solution to the optimization problem (4.70).

Compared to the assumptions in Theorem 4.5 for Douglas-Rachford splitting algorithm, the assumption (4.73) is weaker.

4.4 Proximal Gradient Methods for ℓ^1 Regularization

We here consider an efficient algorithm for ℓ^1 regularization (or LASSO):

$$\underset{\boldsymbol{x}\in\mathbb{R}^n}{\text{minimize}}\ \frac{1}{2}\|\Phi\boldsymbol{x}-\boldsymbol{y}\|_2^2 + \lambda\|\boldsymbol{x}\|_1. \tag{4.74}$$

We assume that $\Phi \in \mathbb{R}^{m\times n}$, $\boldsymbol{y} \in \mathbb{R}^m$, and $\lambda > 0$ are already given.

4.4.1 Algorithm

In (4.74), the first term $\frac{1}{2}\|\Phi\boldsymbol{x}-\boldsymbol{y}\|_2^2$ and the second term $\lambda\|\boldsymbol{x}\|_1$ are both proper, closed, and convex functions of $\boldsymbol{x}$. Also, the proximal operator of the first term, a quadratic function of $\boldsymbol{x}$, is obtained in a closed form as described in Section 4.2.3 (see also Exercise 4.16 below). Hence, we can directly apply the Douglas-Rachford splitting algorithm (4.59) to this problem. Namely, we have the following iterative algorithm based on the Douglas-Rachford splitting algorithm:

$$\begin{aligned}\boldsymbol{x}[k+1] &= \left(\Phi^\top\Phi + \gamma^{-1}I\right)^{-1}\left(\Phi^\top\boldsymbol{y} + \gamma^{-1}\boldsymbol{z}[k]\right),\\ \boldsymbol{z}[k+1] &= \boldsymbol{z}[k] + S_{\gamma\lambda}(2\boldsymbol{x}[k+1]-\boldsymbol{z}[k]) - \boldsymbol{x}[k+1].\end{aligned} \tag{4.75}$$

We note that the proximal operator of $f(\boldsymbol{x}) = \frac{1}{2}\|\Phi\boldsymbol{x}-\boldsymbol{y}\|_2^2$ is given by

$$\mathrm{prox}_{\gamma f}(\boldsymbol{v}) = \left(\Phi^\top\Phi + \gamma^{-1}I\right)^{-1}\left(\Phi^\top\boldsymbol{y} + \gamma^{-1}\boldsymbol{v}\right). \tag{4.76}$$

Exercise 4.16. Prove that the proximal operator of $f(\boldsymbol{x}) = \frac{1}{2}\|\Phi\boldsymbol{x}-\boldsymbol{y}\|_2^2$ is given by (4.76).

As we have seen before, the proximal operator is an "alternative" to the gradient descent update as shown in (4.35). However, the first term $\frac{1}{2}\|\Phi\boldsymbol{x}-\boldsymbol{y}\|_2^2$ of (4.74) is a quadratic function of $\boldsymbol{x}$, which is *differentiable*. By using the gradient of $\frac{1}{2}\|\Phi\boldsymbol{x}-\boldsymbol{y}\|_2^2$ directly, we can develop an efficient algorithm that surpasses the performance of the Douglas-Rachford splitting algorithm. Here we introduce such an algorithm, along with an acceleration method.

First, let us consider the following general problem:

$$\underset{\boldsymbol{x}\in\mathbb{R}^n}{\text{minimize}}\ f_1(\boldsymbol{x})+f_2(\boldsymbol{x}), \tag{4.77}$$

where f_1 is a differentiable and convex function satisfying $\text{dom}(f_1)=\mathbb{R}^n$, and f_2 is a proper, closed, and convex function. Note that f_2 may not be differentiable like the ℓ^1 norm and the indicator function defined in (4.37).

For the optimization problem, we introduce the *proximal gradient algorithm*, which is given as follows:

Proximal gradient algorithm for (4.77)

Initialization: give an initial vector $\boldsymbol{x}[0]$ and a real number $\gamma>0$.
Iteration: for $k=0,1,2,\ldots$ do

$$\boldsymbol{x}[k+1]=\text{prox}_{\gamma f_2}(\boldsymbol{x}[k]-\gamma\nabla f_1(\boldsymbol{x}[k])). \tag{4.78}$$

In this algorithm, $\gamma>0$ is the *step size* of the update. The function $\nabla f_1(\boldsymbol{x})$ is the gradient of f_1 at $\boldsymbol{x}\in\mathbb{R}^n$.

We offer a geometrical interpretation of the proximal gradient algorithm. Let us define

$$\phi(\boldsymbol{x})\triangleq\text{prox}_{\gamma f_2}(\boldsymbol{x}-\gamma\nabla f_1(\boldsymbol{x})). \tag{4.79}$$

Then, from the definition (4.19) of the proximal operator, we have

$$\begin{aligned}\phi(\boldsymbol{x})&=\underset{\boldsymbol{z}\in\mathbb{R}^n}{\arg\min}\left\{f_2(\boldsymbol{z})+\frac{1}{2\gamma}\|\boldsymbol{z}-(\boldsymbol{x}-\gamma\nabla f_1(\boldsymbol{x}))\|_2^2\right\}\\&=\underset{\boldsymbol{z}\in\mathbb{R}^n}{\arg\min}\left\{\tilde{f}_1(\boldsymbol{z};\boldsymbol{x})+f_2(\boldsymbol{z})+\frac{1}{2\gamma}\|\boldsymbol{z}-\boldsymbol{x}\|_2^2\right\},\end{aligned} \tag{4.80}$$

where

$$\tilde{f}_1(\boldsymbol{z};\boldsymbol{x})\triangleq f_1(\boldsymbol{x})+\nabla f_1(\boldsymbol{x})^\top(\boldsymbol{z}-\boldsymbol{x}). \tag{4.81}$$

Note that $\|\nabla f_1(\boldsymbol{x})\|_2^2$ and $f_1(\boldsymbol{x})$ are constant for the minimization with $\boldsymbol{z}$, and hence $\|\nabla f_1(\boldsymbol{x})\|_2^2$ is eliminated and $f_1(\boldsymbol{x})$ is added in (4.80). The function $\tilde{f}_1(\boldsymbol{z};\boldsymbol{x})$ is a linear approximation of $f_1(\boldsymbol{z})$ around the point

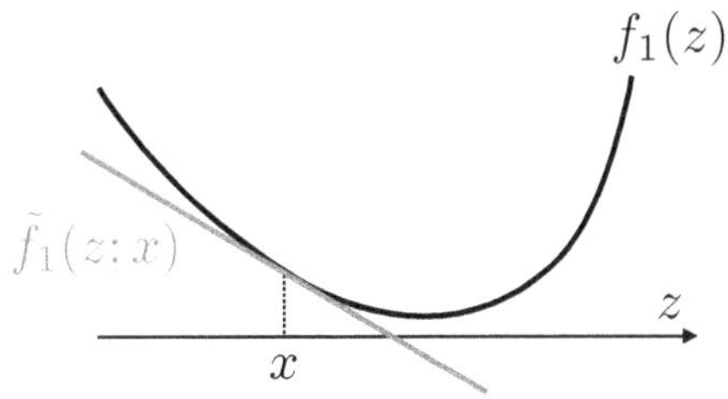

Figure 4.10: Linear approximation $\tilde{f}_1(z;x)$ of convex function $f_1(z)$ at x

$\boldsymbol{x} \in \mathbb{R}^n$. Figure 4.10 shows an example of the linear approximation for one-dimensional case. From (4.80), the function $\phi(\boldsymbol{x})$ is the proximal operator of the linearized function $\tilde{f}_1(\boldsymbol{z};\boldsymbol{x})$ plus $f_2(\boldsymbol{z})$, and the iteration (4.78) can be interpreted as the proximal algorithm (4.24) for this approximated function.

4.4.2 Convergence analysis and acceleration

Here we analyze the convergence of the proximal gradient algorithm. For this, we define Lipschitz continuity. A function $f : \mathbb{R}^n \to \mathbb{R}^n$ is said to be *Lipschitz continuous* over $\mathbb{R}^n$ if there exists a constant $L > 0$ such that the following inequality

$$\|f(\boldsymbol{x}) - f(\boldsymbol{y})\|_2 \leq L\|\boldsymbol{x} - \boldsymbol{y}\|_2 \tag{4.82}$$

holds for any vectors $\boldsymbol{x}, \boldsymbol{y} \in \mathbb{R}^n$. When f is Lipschitz continuous, L is called a *Lipschitz constant*, and the smallest L that satisfies (4.82) is called the *best Lipschitz constant.*

Let us consider the optimization problem (4.77). We assume that the gradient ∇f_1 of f_1 is Lipschitz continuous, that is, there exists $L > 0$ such that the following inequality holds:

$$\|\nabla f_1(\boldsymbol{x}) - \nabla f_1(\boldsymbol{y})\|_2 \leq L\|\boldsymbol{x} - \boldsymbol{y}\|_2, \quad \forall \boldsymbol{x}, \boldsymbol{y} \in \mathbb{R}^n. \tag{4.83}$$

If function f_1 satisfies (4.83), then f_1 is said to be *L-smooth.* Assume that the optimization problem (4.77) has an optimal solution $\boldsymbol{x}^*$. Then we have [122, Section 4.2]

$$\boldsymbol{x}^* = \phi(\boldsymbol{x}^*) = \text{prox}_{\gamma f_2}(\boldsymbol{x}^* - \gamma\nabla f_1(\boldsymbol{x}^*)). \tag{4.84}$$

This implies that an optimal solution $\boldsymbol{x}^*$ to (4.77) is also a *fixed point* of mapping ϕ in (4.79). From this, the meaning of the iteration (4.78) is now clear; this algorithm seeks the fixed point of ϕ.

Exercise 4.17. Consider a continuous function $\phi : \mathbb{R}^n \mapsto \mathbb{R}^n$. Assume that there exists an initial vector $\boldsymbol{x}[0] \in \mathbb{R}^n$ such that the iteration

$$\boldsymbol{x}[k+1] = \phi(\boldsymbol{x}[k]), \quad k = 0, 1, 2, \dots, \tag{4.85}$$

converges to $\boldsymbol{x}^* \in \mathbb{R}^n$. Prove that $\boldsymbol{x}^*$ is a fixed point of the mapping ϕ, that is, $\boldsymbol{x}^* = \phi(\boldsymbol{x}^*)$ holds:

In fact, the following theorem holds [7].

Theorem 4.7. Assume that f_1 is L-smooth, that is, there exists $L > 0$ such that (4.83) holds. Assume also that the step size $\gamma > 0$ satisfies

$$\gamma \leq \frac{1}{L}. \tag{4.86}$$

Then the sequence $\{\boldsymbol{x}[k]\}$ generated by the proximal gradient algorithm (4.78) converges to a solution $\boldsymbol{x}^*$ to (4.77), and we have

$$\|\boldsymbol{x}[k+1] - \boldsymbol{x}^*\|_2 \leq \|\boldsymbol{x}[k] - \boldsymbol{x}^*\|_2, \quad k = 0, 1, 2, \dots \tag{4.87}$$

Moreover, we have

$$f(\boldsymbol{x}[k]) - f(\boldsymbol{x}^*) \leq \frac{L\|\boldsymbol{x}[0] - \boldsymbol{x}^*\|_2^2}{2k}, \quad k = 0, 1, 2, \dots, \tag{4.88}$$

where $f(\boldsymbol{x}) = f_1(\boldsymbol{x}) + f_2(\boldsymbol{x})$.

By this theorem, the convergence rate of the proximal gradient algorithm is $O(1/k)$. Note that this rate is much slower than *linear convergence* (or *first-order convergence*), with which the rate is $O(r^k)$ with $|r| < 1$.

Now, let us derive the proximal gradient algorithm of our ℓ^1 regularization (4.74). In our case, the two functions are

$$f_1(\boldsymbol{x}) = \frac{1}{2}\|\Phi\boldsymbol{x} - \boldsymbol{y}\|_2^2, \quad f_2(\boldsymbol{x}) = \lambda\|\boldsymbol{x}\|_1, \tag{4.89}$$

and the gradient of $f_1(\boldsymbol{x})$ is given by

$$\nabla f_1(\boldsymbol{x}) = \Phi^\top(\Phi\boldsymbol{x} - \boldsymbol{y}). \tag{4.90}$$

Also, the proximal operator of $f_2(\boldsymbol{x}) = \lambda\|\boldsymbol{x}\|_1$ is the soft-thresholding operator (see Section 4.2.5):

$$\mathrm{prox}_{\gamma f_2}(\boldsymbol{v}) = S_{\gamma\lambda}(\boldsymbol{v}). \tag{4.91}$$

From these, the proximal gradient algorithm for (4.74) is given as follows.

Proximal gradient algorithm (ISTA) for (4.74)

Initialization: give an initial vector $\boldsymbol{x}[0]$ and parameter $\gamma > 0$.
Iteration: for $k = 0, 1, 2, \dots$ do

$$\boldsymbol{x}[k+1] = S_{\gamma\lambda}(\boldsymbol{x}[k] - \gamma\Phi^\top(\Phi\boldsymbol{x}[k] - \boldsymbol{y})). \tag{4.92}$$

This algorithm is called the *iterative shrinkage thresholding algorithm*, or *ISTA* for short. From (4.90), a Lipschitz constant of ∇f_1 is given by

$$L = \lambda_{\max}(\Phi^\top\Phi) = \sigma_{\max}(\Phi)^2 = \|\Phi\|^2, \tag{4.93}$$

where $\lambda_{\max}$ and $\sigma_{\max}$ respectively denote the *maximum eigenvalue* and the *maximum singular value*, and $\|\Phi\|$ is a matrix norm defined by $\|\Phi\| \triangleq \sigma_{\max}(\Phi)$. Note that if $\Phi \neq 0$, then $\|\Phi\| > 0$. From (4.86) in Theorem 4.7, if we choose γ to satisfy

$$0 < \gamma \leq \frac{1}{\|\Phi\|^2}, \tag{4.94}$$

then a solution to the ℓ^1 regularization (4.74) is obtained after the simple iteration of (4.92).

Theorem 4.7 implies that the error by ISTA decreases at the rate of $O(1/k)$. We can then accelerate the algorithm by using not only $\boldsymbol{x}[k]$ but also the previous $\boldsymbol{x}[k-1]$ in the k-th step. The following algorithm is the accelerated iteration called *FISTA* (Fast ISTA), which converges at the rate of $O(1/k^2)$ [7], [157].

Fast ISTA (FISTA) for (4.74)

Initialization: give initial vectors $\boldsymbol{x}[0]$, $\boldsymbol{z}[0]$, initial number $t[0]$, and parameter $\gamma > 0$.
Iteration: for $k = 0, 1, 2, \dots$ do

$$\begin{aligned} \boldsymbol{x}[k+1] &= S_{\gamma\lambda}(\boldsymbol{z}[k] - \gamma\Phi^\top(\Phi\boldsymbol{z}[k] - \boldsymbol{y})), \\ t[k+1] &= \frac{1 + \sqrt{1 + 4t[k]^2}}{2}, \\ \boldsymbol{z}[k+1] &= \boldsymbol{x}[k+1] + \frac{t[k]-1}{t[k+1]}(\boldsymbol{x}[k+1] - \boldsymbol{x}[k]). \end{aligned} \tag{4.95}$$

It is surprising that such a simple modification leads to an improvement of computational efficiency from $O(1/k)$ to $(1/k^2)$. However, it is known that $O(1/k^2)$ is optimal and one cannot accelerate the algorithm any further [122, Section 4.3].

4.4.3 Noisy group testing by FISTA

Here we consider the problem of group testing discussed in Section 2.4. We assume the same parameter settings in Section 3.3.2. Namely, the number of individuals is $n = 1000$, among which $s = 5$ individuals are infected. The number of tests is $m = 100$ and the testing matrix $\Phi \in \{0,1\}^{m \times n}$ is chosen randomly. Then, we consider a *noisy* observation, given by

$$\boldsymbol{y} = \Phi \boldsymbol{x} + \boldsymbol{n}, \tag{4.96}$$

where $\boldsymbol{n} \in \mathbb{R}^m$ is a random vector such that each element is independently sampled from the normal distribution with 0 mean and 0.1 variance. From this noisy observation, we reconstruct the original $\boldsymbol{x}$ by solving the ℓ^1 regularization

$$\underset{\boldsymbol{x} \in \mathbb{R}^n}{\text{minimize}} \; \frac{1}{2}\|\Phi \boldsymbol{x} - \boldsymbol{y}\|_2^2 + \lambda \|\boldsymbol{x}\|_1. \tag{4.97}$$

The regularization parameter is $\lambda = 1$.

We solve (4.97) by FISTA in (4.95). We set the initial vectors as $\boldsymbol{x}[0] = \boldsymbol{z}[0] = \boldsymbol{0}$ and the initial number $t[0] = 0$. We choose the step-size parameter $\gamma = 1/\|\Phi\|^2$ to satisfy the inequality (4.94).

Figure 4.11 shows the reconstruction error $\|\boldsymbol{x}_{\text{orig}} - \boldsymbol{x}[k]\|_2$ between the original vector $\boldsymbol{x}_{\text{orig}}$ and the estimated vector $\boldsymbol{x}[k]$, $k = 0, 1, 2, \dots$ by FISTA iteration (4.95). We can see that the error converges to a value after 5000 iterations. We note that the error does not converge to zero due to not only the noise but also the choice of regularization parameter λ. We also note that the error is not monotonically decreasing.

With the iteration, we obtain the estimated vector $\boldsymbol{x}[5000]$, which is shown in Figure 4.12. Since the observation vector $\boldsymbol{y}$ includes noise, the reconstructed vector shows some noise as well. However, the indices of the five largest elements of the estimated vector $\tilde{\boldsymbol{x}}$ match the indices of the non-zero elements of the original vector. Therefore, we can exactly predict the infected 5 individuals from the noisy observation.

The Python program is given below.

```
import cvxpy as cp
import numpy as np
import matplotlib.pyplot as plt
```

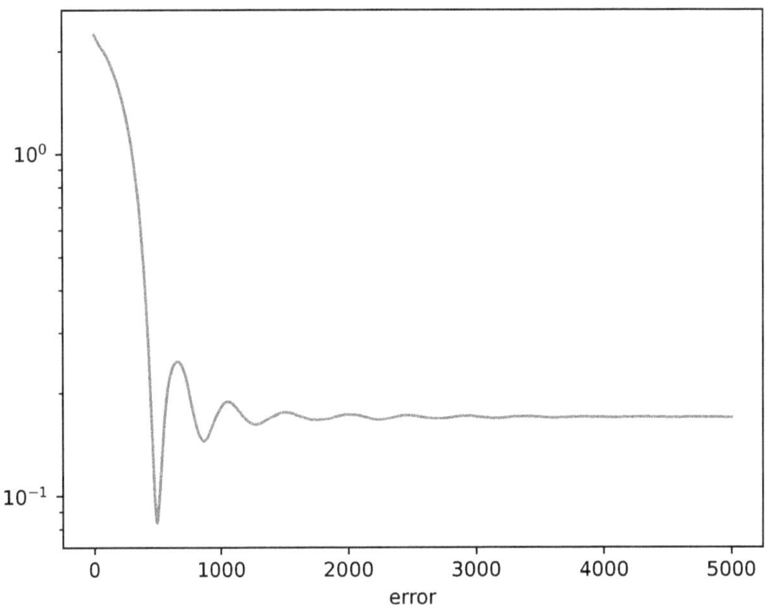

Figure 4.11: The reconstruction error $\|\boldsymbol{x}_{\mathrm{orig}} - \boldsymbol{x}[k]\|_2$ between the original vector $\boldsymbol{x}_{\mathrm{orig}}$ and the estimated vector $\boldsymbol{x}[k]$, $k = 0, 1, 2, \ldots$ by FISTA.

```
## Parameter settings
# Vector size (number of individuals)
n = 1000
# Number of positives
s = 5
# Random seed
np.random.seed(1)
# Original vector (n-dimensional, k-sparse)
x_orig = np.zeros(n)
S = np.random.randint(n,size=s)
x_orig[S] = 1
# Number of tests
m = 100
# Testing matrix
Phi = np.random.randint(2,size=(m, n))
```

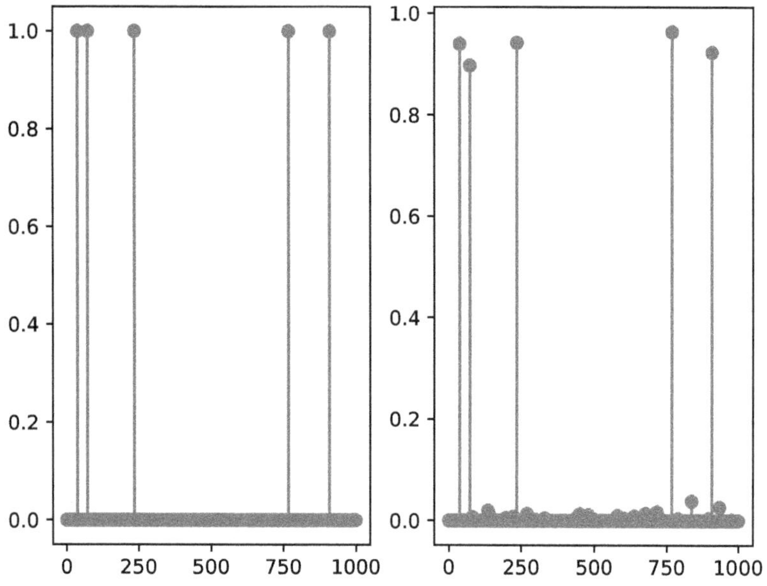

Figure 4.12: The original vector $\boldsymbol{x}_{\text{orig}}$ (left) and the estimated vector $\tilde{\boldsymbol{x}} = \boldsymbol{x}[5000]$ by FISTA.

```
# Result vector
y = Phi @ x_orig + 0.1 * np.random.randn(m)

## Optimization by FISTA
# Soft-thresholding function
def St(lmbd, v):
    n = v.shape[0]
    Sv = np.zeros(n)
    i = np.abs(v) > lmbd
    Sv[i] = v[i] - np.sign(v[i]) * lmbd
    return Sv

# parameter settings
lmbd = 1
Phi_norm = np.linalg.norm(Phi,2)
gamma = 1/Phi_norm**2 # step size
max_itr = 5000 # number of iterations
```

```
x = np.zeros(n) # initial guess for x
z = x # initial guess for z
t = 0 # initial guess for t

error = np.zeros(max_itr) # residual

# FISTA iteration
for k in range(max_itr):
  error[k] = np.linalg.norm(x_orig - x)
  res = Phi @ z - y
  x2 = St(gamma*lmbd, z - gamma*Phi.T @ res)
  t2 = (1 + np.sqrt(1+4*t**2))/2
  z = x2 + (t-1)/t2 * (x2 - x)
  x = x2
  t = t2

## Error plot
fig = plt.figure()
plt.semilogy(error)
plt.xlabel("k")
plt.xlabel("error")

## Reconstructed vector
fig = plt.figure()
ax1 = fig.add_subplot(1, 2, 1)
ax1.stem(x_orig)
ax2 = fig.add_subplot(1, 2, 2)
ax2.stem(x)

## Result
print(np.nonzero(x_orig))
x_est = np.round(x)
print(x_est.nonzero())
```

4.5 Generalized LASSO and ADMM

In this section, we consider an extension of ℓ^1 regularization, with a generalized regularization term:

$$\underset{\boldsymbol{x}\in\mathbb{R}^n}{\text{minimize}}\ \frac{1}{2}\|\Phi\boldsymbol{x}-\boldsymbol{y}\|_2^2+\lambda\|\Psi\boldsymbol{x}\|_1, \tag{4.98}$$

where Ψ is a matrix. We call this optimization problem the *generalized LASSO*. If Ψ is the identity matrix, this problem is reduced to the ℓ^1 regularization, or LASSO, in (4.74). A problem is that the regularization term $\|\Psi\boldsymbol{x}\|_1$ is in general not *proximable*, that is, it is difficult to obtain a closed form of the proximal operator of $\|\Psi\boldsymbol{x}\|_1$. Therefore, we do not directly apply Douglas-Rachford splitting nor the proximal gradient method to this problem. In this section, we introduce an alternative splitting method for this case.

4.5.1 Algorithm

Aside from the generalized LASSO in (4.98), let us consider a general optimization problem:

$$\underset{\boldsymbol{x}\in\mathbb{R}^n,\boldsymbol{z}\in\mathbb{R}^p}{\text{minimize}}\ f_1(\boldsymbol{x})+f_2(\boldsymbol{z})\quad\text{subject to}\quad \boldsymbol{z}=\Psi\boldsymbol{x}, \tag{4.99}$$

where $f_1:\mathbb{R}^n\to\mathbb{R}\cup\{\infty\}$ and $f_2:\mathbb{R}^p\to\mathbb{R}\cup\{\infty\}$ are proper, closed, and convex functions, and $\Psi\in\mathbb{R}^{p\times n}$. The following algorithm, called *Alternating Direction Method of Multipliers*, or *ADMM* for short, is an efficient algorithm to solve (4.99):

ADMM for (4.99)

Initialization: give initial vectors $\boldsymbol{z}[0], \boldsymbol{v}[0]\in\mathbb{R}^p$, and real number $\gamma>0$.

Iteration: for $k=0,1,2,\dots$ do

$$\boldsymbol{x}[k+1]:=\underset{\boldsymbol{x}\in\mathbb{R}^n}{\arg\min}\left\{f_1(\boldsymbol{x})+\frac{1}{2\gamma}\|\Psi\boldsymbol{x}-\boldsymbol{z}[k]+\boldsymbol{v}[k]\|^2\right\}, \tag{4.100}$$

$$\boldsymbol{z}[k+1]:=\text{prox}_{\gamma f_2}(\Psi\boldsymbol{x}[k+1]+\boldsymbol{v}[k]), \tag{4.101}$$

$$\boldsymbol{v}[k+1]:=\boldsymbol{v}[k]+\Psi\boldsymbol{x}[k+1]-\boldsymbol{z}[k+1]. \tag{4.102}$$

To analyze this algorithm, we introduce the *augmented Lagrangian*:

$$L_\rho(\boldsymbol{x},\boldsymbol{z},\boldsymbol{\lambda})=f_1(\boldsymbol{x})+f_2(\boldsymbol{z})+\boldsymbol{\lambda}^\top(\Psi\boldsymbol{x}-\boldsymbol{z})+\frac{\rho}{2}\|\Psi\boldsymbol{x}-\boldsymbol{z}\|_2^2, \tag{4.103}$$

where $\boldsymbol{\lambda}$ is the Lagrange multiplier and ρ is a positive constant. The term 'augmented' means that the function (4.103) is augmented from the usual *Lagrangian*

$$L(\boldsymbol{x}, \boldsymbol{z}, \boldsymbol{\lambda}) = f_1(\boldsymbol{x}) + f_2(\boldsymbol{z}) + \boldsymbol{\lambda}^\top(\Psi\boldsymbol{x} - \boldsymbol{z}), \tag{4.104}$$

by adding the term $\frac{\rho}{2}\|\Psi\boldsymbol{x} - \boldsymbol{z}\|_2^2$. Note that the augmented Lagrangian becomes strongly convex with respect to variables $\boldsymbol{x}$ and $\boldsymbol{z}$ thanks to the additional term $\frac{\rho}{2}\|\Psi\boldsymbol{x} - \boldsymbol{z}\|_2^2$ if $\Psi^\top\Psi$ is positive definite.

Now, let $\gamma \triangleq \rho^{-1}$ and $\boldsymbol{v}[k] \triangleq \gamma\boldsymbol{\lambda}[k]$. Then the ADMM algorithm (4.100)–(4.102) can be rewritten in terms of augmented Lagrangian as

$$\boldsymbol{x}[k+1] = \arg\min_{\boldsymbol{x}\in\mathbb{R}^n} L_\rho(\boldsymbol{x}, \boldsymbol{z}[k], \boldsymbol{\lambda}[k]), \tag{4.105}$$

$$\boldsymbol{z}[k+1] = \arg\min_{\boldsymbol{z}\in\mathbb{R}^p} L_\rho(\boldsymbol{x}[k+1], \boldsymbol{z}, \boldsymbol{\lambda}[k]), \tag{4.106}$$

$$\boldsymbol{\lambda}[k+1] = \boldsymbol{\lambda}[k] + \rho(\Psi\boldsymbol{x}[k+1] - \boldsymbol{z}[k+1]), \quad k = 0, 1, 2, \ldots \tag{4.107}$$

Exercise 4.18. Show that the algorithm in (4.105)–(4.107) is equivalent to (4.100)–(4.102) under the transformation $\gamma = \rho^{-1}$, $\boldsymbol{v}[k] = \gamma\boldsymbol{\lambda}[k]$.

The important point of this algorithm is that the optimization for variables $\boldsymbol{x}$, $\boldsymbol{z}$, and $\boldsymbol{\lambda}$ is decoupled. The first step (4.105) is the minimization of the augmented Lagrangian for the variable $\boldsymbol{x}$ with fixed $\boldsymbol{z}$ and $\boldsymbol{\lambda}$. The second step (4.106) is for $\boldsymbol{z}$ with fixed $\boldsymbol{x}$ and $\boldsymbol{\lambda}$. The third step (4.106) updates the variable $\boldsymbol{\lambda}$ using the previously computed values of $\boldsymbol{x}$ and $\boldsymbol{z}$.

The following is a convergence theorem for the ADMM algorithm [13], [41]:

Theorem 4.8 (Convergence of ADMM). Consider the optimization problem in (4.99). Assume that f_1 and f_2 are proper, closed, and convex functions. Assume also that the Lagrangian (4.104) has a saddle point, that is, there exist $\boldsymbol{x}^*$, $\boldsymbol{z}^*$, and $\boldsymbol{\lambda}^*$ such that

$$L(\boldsymbol{x}^*, \boldsymbol{z}^*, \boldsymbol{\lambda}) \leq L(\boldsymbol{x}^*, \boldsymbol{z}^*, \boldsymbol{\lambda}^*) \leq L(\boldsymbol{x}, \boldsymbol{z}, \boldsymbol{\lambda}^*), \quad \forall \boldsymbol{x}, \boldsymbol{z}, \boldsymbol{\lambda}. \tag{4.108}$$

Then, the ADMM algorithm (4.100)–(4.102) satisfies the following convergence properties:

- The residual

$$\boldsymbol{r}[k] \triangleq \Psi\boldsymbol{x}[k] - \boldsymbol{z}[k], \quad k = 0, 1, 2, \ldots, \tag{4.109}$$

converges to $\mathbf{0}$ as $k \to \infty$. This implies that the iterates converge to a feasible solution to (4.99).

- The objective value $f_1(\boldsymbol{x}[k])+f_2(\boldsymbol{z}[k])$ converges to the optimal value
$$f^* \triangleq \inf_{\substack{\boldsymbol{x}\in\mathbb{R}^n,\boldsymbol{z}\in\mathbb{R}^p\\ \Psi\boldsymbol{x}=\boldsymbol{z}}} f_1(\boldsymbol{x}) + f_2(\boldsymbol{z}). \tag{4.110}$$
- If $\Psi^\top\Psi$ is positive definite, then the sequence $\{(\boldsymbol{x}[k], \boldsymbol{z}[k])\}$ converges to an optimal solution $(\boldsymbol{x}^*, \boldsymbol{z}^*)$ to the optimization problem (4.99).

We can now derive the ADMM algorithm for the generalized LASSO (4.98). First, since $f_1(\boldsymbol{x}) = \frac{1}{2}\|\Phi\boldsymbol{x} - \boldsymbol{y}\|_2^2$, the minimization in (4.100) becomes

$$\begin{aligned}\arg\min_{\boldsymbol{x}\in\mathbb{R}^n} &\left\{\frac{1}{2}\|\Phi\boldsymbol{x} - \boldsymbol{y}\|_2^2 + \frac{1}{2\gamma}\|\Psi\boldsymbol{x} - \boldsymbol{z}[k] + \boldsymbol{v}[k]\|_2^2\right\}\\ &= \left(\Phi^\top\Phi + \frac{1}{\gamma}\Psi^\top\Psi\right)^{-1}\left(\Phi^\top\boldsymbol{y} + \frac{1}{\gamma}\Psi^\top(\boldsymbol{z}[k] - \boldsymbol{v}[k])\right).\end{aligned} \tag{4.111}$$

Exercise 4.19. Prove the equality in (4.111).

Next, since $f_2(\boldsymbol{x}) = \lambda\|\boldsymbol{x}\|_1$, the proximal operator in (4.101) is the soft-thresholding function. In summary, the ADMM algorithm for the generalized LASSO (4.98) is given as follows.

ADMM for generalized LASSO (4.98)

Initialization: give initial vectors $\boldsymbol{z}[0], \boldsymbol{v}[0] \in \mathbb{R}^p$, and real number $\gamma > 0$.
Iteration: for $k = 0, 1, 2, \ldots$ do

$$\boldsymbol{x}[k+1] = \left(\Phi^\top\Phi + \frac{1}{\gamma}\Psi^\top\Psi\right)^{-1}\left(\Phi^\top\boldsymbol{y} + \frac{1}{\gamma}\Psi^\top(\boldsymbol{z}[k] - \boldsymbol{v}[k])\right), \tag{4.112}$$

$$\boldsymbol{z}[k+1] = S_{\gamma\lambda}(\Psi\boldsymbol{x}[k+1] + \boldsymbol{v}[k]), \tag{4.113}$$

$$\boldsymbol{v}[k+1] = \boldsymbol{v}[k] + \Psi\boldsymbol{x}[k+1] - \boldsymbol{z}[k+1]. \tag{4.114}$$

If we compute the inverse matrix $(\Phi^\top\Phi+\gamma^{-1}\Psi^\top\Psi)^{-1}$ *offline* (i.e., before the iteration), the above ADMM algorithm just includes matrix-vector multiplication, vector addition, and element-wise soft-thresholding. By this property, one can implement this algorithm in a small device and execute very fast. Moreover, if the matrix $\Phi^\top\Phi + \gamma^{-1}\Psi^\top\Psi$ is a *tridiagonal matrix*, the linear equation

$$\left(\Phi^\top \Phi + \frac{1}{\gamma} \Psi^\top \Psi\right) \boldsymbol{x} = \boldsymbol{v} \tag{4.115}$$

with unknown $\boldsymbol{x}$ can be solved in $O(n)$ [48, Section 4.3], and the first step (4.112) can be computed very efficiently.

4.5.2 Total variation denoising

Here we consider *total variation denoising* for images, which can achieve noise reduction and edge preserving at the same time. Let us assume that we have a noisy image $\boldsymbol{Y} \in \mathbb{R}^{n \times m}$, where each element in $\boldsymbol{Y}$ is the pixel value of the image of size $n \times m$. From 2D image data $\boldsymbol{Y}$, we pull out each column vector, say $\boldsymbol{y} \in \mathbb{R}^n$, and solve the following optimization problem, one by one:

$$\underset{\boldsymbol{x} \in \mathbb{R}^n}{\text{minimize}} \; \|\boldsymbol{x} - \boldsymbol{y}\|_2^2 + \lambda \sum_{i=1}^{n-1} |x_{i+1} - x_i|. \tag{4.116}$$

The first term is the ℓ^2 error between $\boldsymbol{x}$ and $\boldsymbol{y}$ for proximity to the data, while the second term is the ℓ^1 norm of the difference, called the *total variation*, for flatness of the result. The optimization problem (4.116) is a special case of the generalized LASSO (4.98) with $\Phi = I$ (identity matrix) and

$$\Psi = \begin{bmatrix} -1 & 1 & 0 & \dots & 0 \\ 0 & -1 & 1 & \ddots & \vdots \\ \vdots & \ddots & \ddots & \ddots & 0 \\ 0 & \dots & 0 & -1 & 1 \end{bmatrix}. \tag{4.117}$$

We can directly exploit ADMM (4.112)–(4.114) for this problem. Moreover, the matrix $\Phi^\top \Phi + \gamma^{-1} \Psi^\top \Psi$ is a tridiagonal matrix, and the algorithm can be executed very fast, as mentioned above.

The total variation, which is described by $\|\Psi \boldsymbol{x}\|_1$, can be explained as a convex approximation of the ℓ^0 total variation $\|\Psi \boldsymbol{x}\|_0$. Minimizing the ℓ^0 total variation leads to a sparse difference vector, and hence the optimization result can be maximally flat (i.e., the difference $= 0$ in all but a few pixels). A few nonzero differences come from image edges. That is, we assume that there are just a few edges in an image.

Now, we show the results of total variation denoising. Figure 4.13 shows the original image and a noisy image. The noise in the noisy image is so-called *salt-and-pepper noise* with noise density 0.05. Roughly speaking,

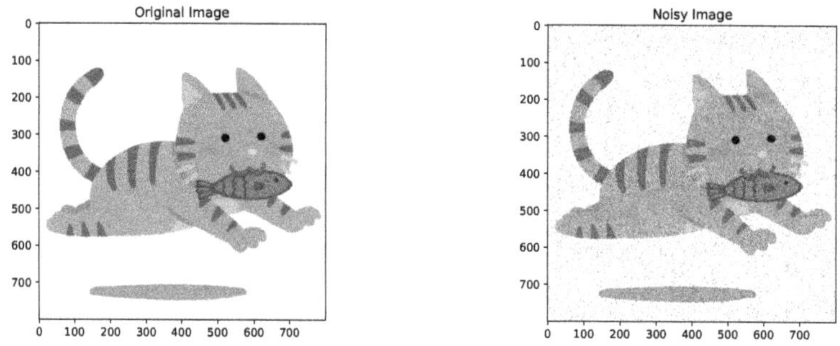

Figure 4.13: Original image (left) and noisy image (right)

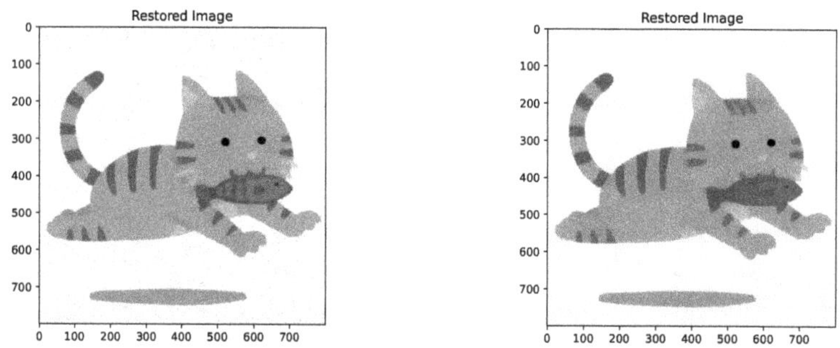

Figure 4.14: Total variation denoising with $\lambda = 50$ (left), $\lambda = 100$ (right)

about 5% of the original pixels are randomly replaced by black or white pixels.

From the noisy image, we remove noise by the total variation denoising. We use the ADMM algorithm with $\gamma = 1$. The maximum number of iterations in ADMM is set to $N = 500$. For the 2-D image, we first run total variation denoising horizontally and then vertically. That is, we run the algorithm twice for one image. Figure 4.14 shows the results of denoising with $\lambda = 50$ and $\lambda = 100$. If you take larger λ, the variation between adjacent pixels will be smaller. Comparing images with $\lambda = 50$ and $\lambda = 100$ in Figure 4.14, the image with $\lambda = 100$ gives an impression of more smoothness than that with $\lambda = 50$. This effect is much more perceptible when $\lambda = 200$. Figure 4.15 shows the result. The total variation term is too strong in this case, and the restored image is now unacceptable.

In summary, to obtain a good result, you should carefully choose the parameter λ, which affects the quality of denoising. However, there is no

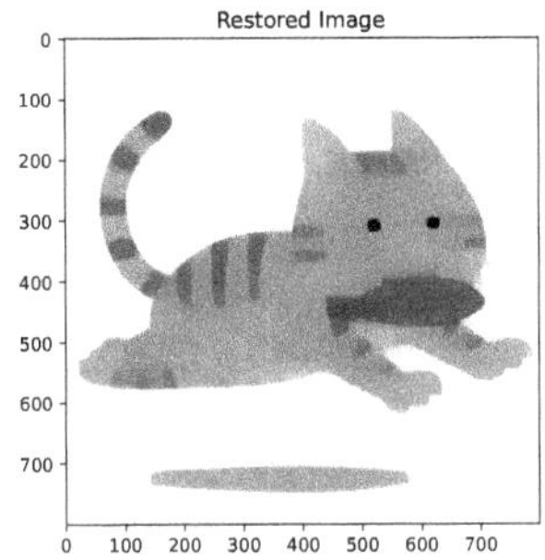

Figure 4.15: Result of total variation denoising with $\lambda = 200$

general rule for optimal λ, and you should choose λ by trial and error.

A Python program for this simulation is given below. You can experiment with the total variation denoising by yourself.

```
import numpy as np
from PIL import Image
import matplotlib.pyplot as plt
from scipy.sparse import eye, diags, kron
from scipy.sparse.linalg import spsolve

## Settings
# Soft-thresholding function
def soft_thresholding(lam, X):
    return np.sign(X) * np.maximum(np.abs(X) - lam
        , 0)

# Read image and convert to grayscale
img = Image.open('cat.jpg').convert('L')
X_orig = np.array(img)
n, m = X_orig.shape

# Add salt and pepper noise
np.random.seed(1)
Y = X_orig.copy()
num_salt = np.ceil(0.05 * Y.size * 0.5)
coords = [np.random.randint(0, i - 1, int(num_salt
   )) for i in Y.shape]
```

```
Y[coords[0], coords[1]] = 255
Y[coords[1], coords[0]] = 0

# Display original and noisy images
plt.figure()
plt.imshow(X_orig, cmap='gray')
plt.title('Original Image')
plt.show()

plt.figure()
plt.imshow(Y, cmap='gray')
plt.title('Noisy Image')
plt.show

## Total variation denoising
# Denoising parameters
lambda_val = 50
gamma = 1
N = 500
Phi = eye(n)
Psi = -diags([np.ones(n)], [0]) + diags([np.ones(n
    - 1)], [1])

# Matrix M
M = (Phi.T * Phi + (1 / gamma) * (Psi.T * Psi)).
    tocsc()

# Optimization by ADMM
X_res = np.zeros_like(Y, dtype=float)
Z = np.zeros_like(Y, dtype=float)
V = np.zeros_like(Y, dtype=float)
W = Phi.T @ Y.astype(float)
for k in range(N):
    X_res = spsolve(M, W + gamma * Psi.T @ (Z - V)
        )
    P = Psi @ X_res + V
    Z = soft_thresholding(gamma * lambda_val, P)
    V = P - Z
```

```

# Horizontal processing
W = np.rot90(X_res)
for k in range(N):
    X_res = spsolve(M, W + gamma * Psi.T @ (Z - V)
        )
    P = Psi @ X_res + V
    Z = soft_thresholding(gamma * lambda_val , P)
    V = P - Z
X_res = np.rot90(X_res , -1)

## Show the restored image
plt.figure()
plt.imshow(np.clip(X_res.round(), 0, 255).astype(
    np.uint8), cmap='gray')
plt.title('Restored Image')
plt.show()
```

4.6 Further Readings

To study convex optimization deeply, you can choose a renowned book by Boyd and Vandenberghe [14]. The PDF version of the book, lecture slides, and Python programs for exercises can be available in
`http://web.stanford.edu/~boyd/cvxbook/`
You can also choose a recent book by Bertsekas [9]. This book devotes much space to recent algorithms such as the proximal gradient algorithms and ADMM. If you need deep and mathematical knowledge of convex optimization at a research level, you can consult the book [6] by Bauschke and Combettes. For proximal splitting algorithms, you can refer to [29], [122]. The book chapter [7] by Beck and Teboulle is a nice introduction to ISTA and FISTA. For ADMM, you can read the book [13] by Boyd et al.

Chapter 5

Greedy Algorithms

In the previous chapter, we have formulated the problem of sparse representation as optimization problems with ℓ^1 norm, for which there are efficient and fast algorithms. The idea was to approximate the non-convex and discontinuous ℓ^0 norm by the convex ℓ^1 norm. In this chapter, we explore alternative algorithms that directly solve ℓ^0-norm optimization problems by using greedy methods.

Key ideas of Chapter 5

- Greedy algorithms are available to directly solve ℓ^0 optimization.
- The greedy algorithms introduced in this chapter show the linear convergence, which are much faster than the proximal splitting algorithms.
- A local optimal solution is obtained by a greedy algorithm, which is not necessarily a global optimizer.

5.1 ℓ^0 Optimization

Let us consider the following ℓ^0 optimization problem:

$$\underset{\boldsymbol{x}\in\mathbb{R}^n}{\text{minimize}} \quad \|\boldsymbol{x}\|_0 \quad \text{subject to} \quad \Phi\boldsymbol{x} = \boldsymbol{y}, \tag{5.1}$$

where we assume $\Phi \in \mathbb{R}^{m\times n}$ and $\boldsymbol{y} \in \mathbb{R}^m$ are given. To consider this optimization problem, let us first define the *mutual coherence* of a matrix.

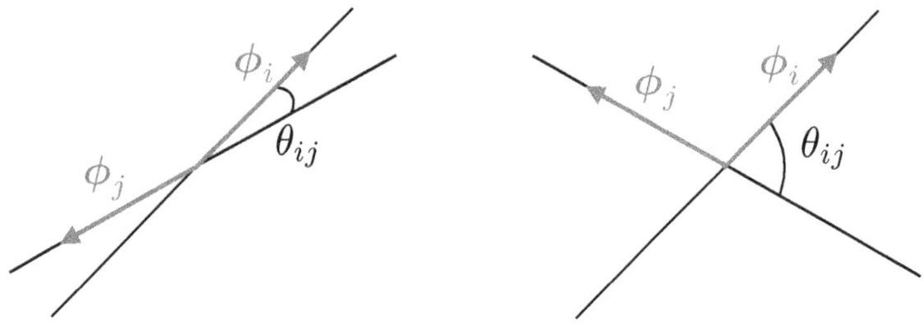

Figure 5.1: Angle θ_{ij} between two lines along with ϕ_i and ϕ_j: coherent vectors (left) and incoherent vectors (right)

Definition 5.1. For a matrix $\Phi = [\phi_1, \phi_2, \ldots, \phi_n] \in \mathbb{R}^{m\times n}$ with $\phi_i \in \mathbb{R}^m$, $i = 1, 2, \ldots, n$, we define the *mutual coherence* $\mu(\Phi)$ by

$$\mu(\Phi) \triangleq \max_{\substack{i,j=1,\ldots,n\\ i\neq j}} \frac{|\langle \phi_i, \phi_j\rangle|}{\|\phi_i\|_2\|\phi_j\|_2}. \tag{5.2}$$

The mutual coherence is the maximum value of the absolute value of the inner product of $\phi_i/\|\phi_i\|_2$ and $\phi_j/\|\phi_j\|_2$. That is,

$$\mu(\Phi) = \max_{\substack{i,j=1,\ldots,n\\ i\neq j}} \left|\left\langle \frac{\phi_i}{\|\phi_i\|_2}, \frac{\phi_j}{\|\phi_j\|_2}\right\rangle\right|. \tag{5.3}$$

The value $\left\langle \frac{\phi_i}{\|\phi_i\|_2}, \frac{\phi_j}{\|\phi_j\|_2}\right\rangle$ is the *cosine* of the angle θ_{ij} between two lines along with ϕ_i and ϕ_j. If the angle is small (i.e. coherent), then this value is large (close to 1), and if the angle is large (close to 90°, incoherent), then the value is almost 0. Figure 5.1 illustrates these properties. Hence, the mutual coherence is described as

$$\mu(\Phi) = \max_{\substack{i,j=1,\ldots,n\\ i\neq j}} |\cos\theta_{ij}|. \tag{5.4}$$

Roughly speaking, if the vectors $\phi_1, \ldots, \phi_n$ are uniformly spread in $\mathbb{R}^m$, then the mutual coherence $\mu(\Phi)$ is small. On the other hand, if some vectors in Φ are coherent like a tassel, then $\mu(\Phi)$ is large. Figure 5.2 shows examples of dictionaries $\{\phi_1, \phi_2, \phi_3\}$ with both large and small values of $\mu(\Phi)$.

From Cauchy-Schwartz inequality

$$|\langle \boldsymbol{x}, \boldsymbol{y}\rangle| \leq \|\boldsymbol{x}\|_2\|\boldsymbol{y}\|_2, \quad \forall \boldsymbol{x}, \boldsymbol{y} \in \mathbb{R}^m, \tag{5.5}$$

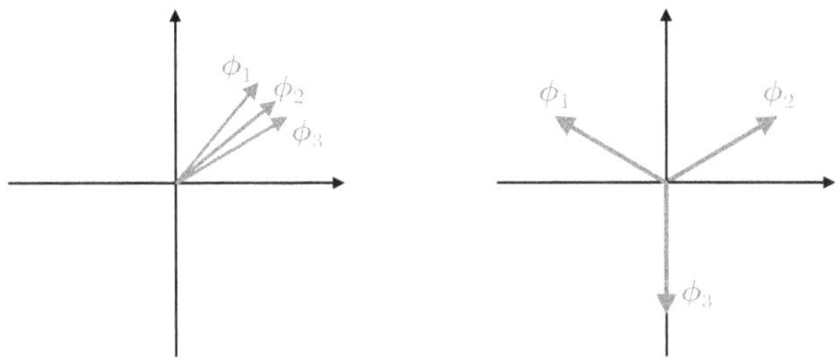

Figure 5.2: Dictionary $\{\phi_1, \phi_2, \phi_3\}$ with large $\mu(\Phi)$ (left) and small $\mu(\Phi)$ (right)

the maximum value of the mutual coherence is 1. Since the equality in (5.5) holds if and only if the two vectors $\boldsymbol{x}$ and $\boldsymbol{y}$ are parallel, we have $\mu(\Phi) = 1$ if and only if there exist parallel vectors in $\{\phi_1, \phi_2, \ldots, \phi_n\}$. On the other hand, the mutual coherence is always non-negative, and $\mu(\Phi) = 0$ if $\phi_1, \phi_2, \ldots, \phi_n$ are mutually orthogonal.

By using the mutual coherence, we can characterize the solution of the ℓ^0 optimization (5.1) [43, Theorem 2.5]:

Theorem 5.1. Suppose $\boldsymbol{y} \neq \boldsymbol{0}$. If there exists a vector $\boldsymbol{x} \in \mathbb{R}^n$ that satisfies the linear equation $\Phi \boldsymbol{x} = \boldsymbol{y}$, and

$$\|\boldsymbol{x}\|_0 < \frac{1}{2}\left(1 + \frac{1}{\mu(\Phi)}\right), \tag{5.6}$$

then $\boldsymbol{x}$ is the sparsest solution of the linear equation.

By this theorem, let us consider properties of the solution(s) of the ℓ^0 optimization problem in (5.1).

First, let us assume $\mu(\Phi) < 1$. That is, there are no parallel vectors in $\{\phi_1, \phi_2, \ldots, \phi_n\}$. Then, we have

$$\frac{1}{2}\left(1 + \frac{1}{\mu(\Phi)}\right) > 1, \tag{5.7}$$

and hence if there exists a 1-sparse solution $\boldsymbol{x}$ (i.e. $\|\boldsymbol{x}\|_0 = 1$) of equation $\Phi \boldsymbol{x} = \boldsymbol{y}$, then this is the sparsest solution from Theorem 5.1. Now, we have

$$\boldsymbol{y} = \Phi \boldsymbol{x} = x_1 \phi_1 + x_2 \phi_2 + \cdots + x_n \phi_n, \tag{5.8}$$

and hence the 1-sparse solution is parallel to one of $\phi_1, \phi_2, \ldots, \phi_n$. From this, we find ϕ_i that is parallel to $\boldsymbol{y}$. This is formulated as a problem to find an index $i \in \{1, 2, \ldots, n\}$ that minimizes the error $e(i)$ defined by

$$e(i) \triangleq \min_{x \in \mathbb{R}} \|x\boldsymbol{\phi}_i - \boldsymbol{y}\|_2^2. \tag{5.9}$$

If there exists $\boldsymbol{x}$ with $\|\boldsymbol{x}\|_0 = 1$, then there exists an index $i \in \{1, 2, \ldots, n\}$ that achieves $e(i) = 0$. Now, the minimum value of (5.9) can be easily obtained as follows:

$$\begin{aligned} e(i) &= \min_{x \in \mathbb{R}} \|x\boldsymbol{\phi}_i - \boldsymbol{y}\|_2^2 \\ &= \min_{x \in \mathbb{R}} \left\{ \langle \boldsymbol{\phi}_i, \boldsymbol{\phi}_i \rangle x^2 - 2\langle \boldsymbol{\phi}_i, \boldsymbol{y} \rangle x + \langle \boldsymbol{y}, \boldsymbol{y} \rangle \right\} \\ &= \min_{x \in \mathbb{R}} \left\{ \|\boldsymbol{\phi}_i\|_2^2 \left(x - \frac{\langle \boldsymbol{\phi}_i, \boldsymbol{y} \rangle}{\|\boldsymbol{\phi}_i\|_2^2} \right)^2 + \|\boldsymbol{y}\|_2^2 - \frac{\langle \boldsymbol{\phi}_i, \boldsymbol{y} \rangle^2}{\|\boldsymbol{\phi}_i\|_2^2} \right\} \\ &= \|\boldsymbol{y}\|_2^2 - \frac{\langle \boldsymbol{\phi}_i, \boldsymbol{y} \rangle^2}{\|\boldsymbol{\phi}_i\|_2^2}. \end{aligned} \tag{5.10}$$

From this formula, we can find one index i^* that satisfies $e(i^*) = 0$ (if it exists) by computing $e(i)$ for $i = 1, 2, \ldots, n$. Then we have

$$\boldsymbol{y} = x^* \boldsymbol{\phi}_{i^*}, \quad x^* \triangleq \frac{\langle \boldsymbol{\phi}_{i^*}, \boldsymbol{y} \rangle}{\|\boldsymbol{\phi}_{i^*}\|_2^2}, \tag{5.11}$$

and the corresponding 1-sparse vector $\boldsymbol{x}^*$ is given by

$$\boldsymbol{x}^* = [0, \ldots, 0, \overset{\overset{i^*}{\vee}}{x^*}, 0, \ldots, 0]^\top. \tag{5.12}$$

This computation requires $O(n)$ computational time at the worst case.

Let us generalize this observation. Assume that there exists a natural number k that satisfies

$$\mu(\Phi) < \frac{1}{2k - 1}. \tag{5.13}$$

Then we have

$$\frac{1}{2}\left(1 + \frac{1}{\mu(\Phi)}\right) > \frac{1}{2}(1 + 2k - 1) = k. \tag{5.14}$$

Assume also that there exists a k-sparse solution (i.e. $\|\boldsymbol{x}\|_0 \leq k$) of the linear equation $\Phi\boldsymbol{x} = \boldsymbol{y}$. From Theorem 5.1, this is the sparsest solution. Then, the vector $\boldsymbol{y}$ is a linear combination of k vectors in the dictionary $\{\boldsymbol{\phi}_1, \boldsymbol{\phi}_2, \ldots, \boldsymbol{\phi}_n\}$. As we have seen in Section 2.5 in Chapter 2 (p. 27), to find the k-sparse solution by the exhaustive search, we need $\binom{n}{k}$ or $O(n^k)$ computations, which cannot be acceptable in large-scale problems.

For such problems, a method called the *greedy method* is available. This method is an iterative method for a global solution, in which the locally optimal choice is made at each stage. Although this method does not always give a global solution, this method is a powerful tool for combinatorial problems. In the next section, we introduce greedy algorithms for the ℓ^0 optimization problem in (5.1).

5.2 Orthogonal Matching Pursuit

5.2.1 Matching pursuit (MP)

First, we introduce the simplest greedy algorithm called *matching pursuit* (MP for short) to solve the ℓ^0 optimization problem in (5.1). This algorithm iteratively seeks a 1-sparse vector that is a solution of a local ℓ^0 optimization problem. As mentioned above, a 1-sparse optimal vector is easily obtained with $O(n)$ computations. Matching pursuit aims at finding an optimal solution by iteratively solving a series of simpler local optimization problems.

The algorithm of matching pursuit iteratively approximates the solution of linear equation $\Phi \boldsymbol{x} = \boldsymbol{y}$ by decreasing the *residual* $\boldsymbol{r}[k] = \boldsymbol{y} - \Phi \boldsymbol{x}[k]$ at each step. The procedure is shown as follows:

1. Find a 1-sparse vector $\boldsymbol{x}[1]$ that minimizes $\|\Phi \boldsymbol{x} - \boldsymbol{y}\|_2$.

2. For $k = 1, 2, 3, \ldots$ do

   - Compute the residual $\boldsymbol{r}[k] = \boldsymbol{y} - \Phi \boldsymbol{x}[k]$.
   - Find a 1-sparse vector $\boldsymbol{x}^*$ that minimizes $\|\Phi \boldsymbol{x} - \boldsymbol{r}[k]\|_2$ and set

$$\boldsymbol{x}[k+1] = \boldsymbol{x}[k] + \boldsymbol{x}^*.$$

At the first step, we seek a 1-sparse vector $\boldsymbol{x}[1]$ that minimizes $\|\Phi \boldsymbol{x} - \boldsymbol{y}\|_2$. Let $x[1]$ be the non-zero element of $\boldsymbol{x}[1]$ and $i[k]$ the corresponding index, that is,

$$\boldsymbol{x}[1] = [0, \ldots, 0, \overset{\overset{i[1]}{\vee}}{x[1]}, 0, \ldots, 0]^\top = x[1]\boldsymbol{e}_{i[1]}, \tag{5.15}$$

where $\boldsymbol{e}_i$, $i \in \{1, \ldots, n\}$ is the standard basis in $\mathbb{R}^n$ defined by

$$\boldsymbol{e}_i \triangleq [0, \ldots, 0, \overset{\overset{i}{\vee}}{1}, 0, \ldots, 0]^\top \in \mathbb{R}^n. \tag{5.16}$$

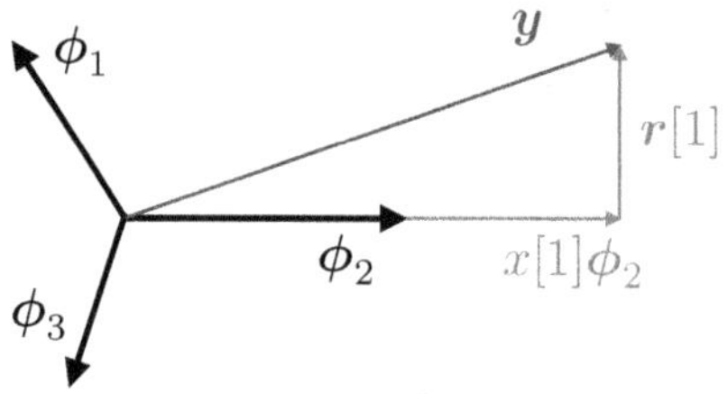

Figure 5.3: Vector $\tilde{\boldsymbol{y}}[1] = x[1]\boldsymbol{\phi}_2$ with $i[1] = 2$ that is the best 1-sparse approximation of $\boldsymbol{y}$. The residual vector $\boldsymbol{r}[1]$ is orthogonal to $\boldsymbol{\phi}_2$.

Then, from (5.10), $i[1]$ and $x[1]$ are easily obtained as

$$\begin{aligned} i[1] &= \underset{i\in\{1,\dots,n\}}{\arg\min}\; e(i) \\ &= \underset{i\in\{1,\dots,n\}}{\arg\min} \left\{ \|\boldsymbol{y}\|_2^2 - \frac{\langle \boldsymbol{\phi}_i, \boldsymbol{y}\rangle^2}{\|\boldsymbol{\phi}_i\|_2^2} \right\} \\ &= \underset{i\in\{1,\dots,n\}}{\arg\max}\; \frac{\langle \boldsymbol{\phi}_i, \boldsymbol{y}\rangle^2}{\|\boldsymbol{\phi}_i\|_2^2}, \\ x[1] &= \frac{\langle \boldsymbol{\phi}_{i[1]}, \boldsymbol{y}\rangle}{\|\boldsymbol{\phi}_{i[1]}\|_2^2}. \end{aligned} \tag{5.17}$$

The residual is given by

$$\boldsymbol{r}[1] = \boldsymbol{y} - \Phi\boldsymbol{x}[1] = \boldsymbol{y} - x[1]\boldsymbol{\phi}_{i[1]}, \tag{5.18}$$

and we have

$$\boldsymbol{y} = x[1]\boldsymbol{\phi}_{i[1]} + \boldsymbol{r}[1]. \tag{5.19}$$

We can easily check (see Exercise 5.1 below) that the residual vector $\boldsymbol{r}[1]$ is orthogonal to $\boldsymbol{\phi}_{i[1]}$, and hence we have

$$\|\boldsymbol{y}\|_2^2 = \|x[1]\boldsymbol{\phi}_{i[1]}\|_2^2 + \|\boldsymbol{r}[1]\|_2^2. \tag{5.20}$$

If the residual $\|\boldsymbol{r}[1]\|_2$ is sufficiently small, then

$$\tilde{\boldsymbol{y}}[1] \triangleq x[1]\boldsymbol{\phi}_{i[1]} = \Phi\boldsymbol{x}[1] \tag{5.21}$$

is a good approximation of $\boldsymbol{y}$. Figure 5.3 illustrates this observation.

Exercise 5.1. Prove that the residual vector $\boldsymbol{r}[1]$ is orthogonal to $\boldsymbol{\phi}_{i[1]}$. Also prove that the equation (5.20) holds.

At the second step, we seek a 1-sparse vector that is the best approximation of the residual vector $\boldsymbol{r}[1]$ in (5.18). The 1-sparse vector is easily

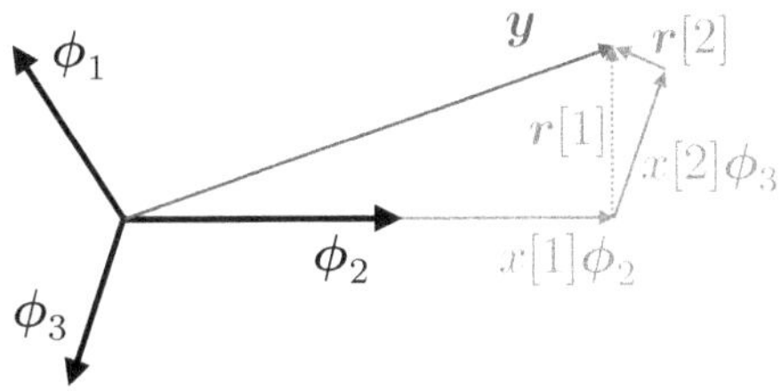

Figure 5.4: Vector $x[2]\phi_3$ with $i[2] = 3$ that is the best 1-sparse approximation of the residual vector $\boldsymbol{r}[1]$. The residual vector $\boldsymbol{r}[2]$ is orthogonal to $\boldsymbol{\phi}_3$, and $\boldsymbol{y} = x[1]\boldsymbol{\phi}_2 + x[2]\boldsymbol{\phi}_3 + \boldsymbol{r}[2]$ holds.

obtained by (5.10) with $\boldsymbol{r}[1]$ instead of $\boldsymbol{y}$. Let $\boldsymbol{x}[2]$ be the 1-sparse vector, $x[2]$ its non-zero element, and $i[2]$ the corresponding index. Then we have

$$i[2] = \underset{i \in \{1,\dots,n\}}{\arg\max} \frac{\langle \boldsymbol{\phi}_i, \boldsymbol{r}[1] \rangle^2}{\|\boldsymbol{\phi}_i\|_2^2}, \quad x[2] = \frac{\langle \boldsymbol{\phi}_{i[2]}, \boldsymbol{r}[1] \rangle}{\|\boldsymbol{\phi}_{i[2]}\|_2^2}. \tag{5.22}$$

The residual vector $\boldsymbol{r}[2]$ is given by

$$\boldsymbol{r}[2] = \boldsymbol{r}[1] - \Phi \boldsymbol{x}[2] = \boldsymbol{r}[1] - x[2]\boldsymbol{\phi}_{i[2]}, \tag{5.23}$$

and from (5.19), we have

$$\boldsymbol{y} = x[1]\boldsymbol{\phi}_{i[1]} + x[2]\boldsymbol{\phi}_{i[2]} + \boldsymbol{r}[2]. \tag{5.24}$$

It is easily shown that $\boldsymbol{\phi}_{i[2]}$ and $\boldsymbol{r}[2]$ are orthogonal to each other, and

$$\|\boldsymbol{r}[1]\|_2^2 = \|x[2]\boldsymbol{\phi}_{i[2]}\|_2^2 + \|\boldsymbol{r}[2]\|_2^2. \tag{5.25}$$

From this with (5.20), we have

$$\|\boldsymbol{y}\|_2^2 = \|x[1]\boldsymbol{\phi}_{i[1]}\|_2^2 + \|x[2]\boldsymbol{\phi}_{i[2]}\|_2^2 + \|\boldsymbol{r}[2]\|_2^2. \tag{5.26}$$

Now we obtain a 2-sparse vector

$$\boldsymbol{x}[2] \triangleq [0, \dots, 0, \underset{\substack{\vee \\ i[1]}}{x[1]}, 0, \dots, 0, \underset{\substack{\vee \\ i[2]}}{x[2]}, 0, \dots, 0]^\top = x[1]\boldsymbol{e}_{i[1]} + x[2]\boldsymbol{e}_{i[2]}, \tag{5.27}$$

which gives a 2-sparse approximation of $\boldsymbol{y}$ as

$$\tilde{\boldsymbol{y}}[2] \triangleq x[1]\boldsymbol{\phi}_{i[1]} + x[2]\boldsymbol{\phi}_{i[2]} = \Phi \boldsymbol{x}[2]. \tag{5.28}$$

Figure 5.4 illustrates this.

If we continue the same procedure, we have the following equation at the k-th step:

$$\boldsymbol{y} = x[1]\boldsymbol{\phi}_{i[1]} + x[2]\boldsymbol{\phi}_{i[2]} + \cdots + x[k]\boldsymbol{\phi}_{i[k]} + \boldsymbol{r}[k]. \tag{5.29}$$

Define the k-sparse vector by

$$\boldsymbol{x}[k] \triangleq x[1]\boldsymbol{e}_{i[1]} + x[2]\boldsymbol{e}_{i[2]} + \cdots + x[k]\boldsymbol{e}_{i[k]}. \tag{5.30}$$

Then the vector $\boldsymbol{y}$ is approximated by using this k-sparse vector as

$$\tilde{\boldsymbol{y}}[k] \triangleq x[1]\boldsymbol{\phi}_{i[1]} + x[2]\boldsymbol{\phi}_{i[2]} + \cdots + x[k]\boldsymbol{\phi}_{i[k]} = \Phi\boldsymbol{x}[k]. \tag{5.31}$$

If the residual $\boldsymbol{r}[k]$ is sufficiently small, we can terminate the algorithm at iteration k. This yields a k-sparse approximated solution to the ℓ^0 optimization problem in (5.1). The exhaustive search requires $O(n^k)$ computations to obtain a k-sparse vector, while matching pursuit needs just $O(nk)$ computations.

We summarize the algorithm of matching pursuit as follows:

MP for ℓ^0 optimization (5.1)

Initialization: set $\boldsymbol{x}[0] = \boldsymbol{0}$, $\boldsymbol{r}[0] = \boldsymbol{y}$, and $k = 1$.
Iteration: while $\|\boldsymbol{r}[k]\|_2 \geq \mathsf{eps}$, do

$$\begin{aligned} i[k] &:= \operatorname*{arg\,max}_{i\in\{1,\ldots,n\}} \frac{\langle \boldsymbol{\phi}_i, \boldsymbol{r}[k-1]\rangle^2}{\|\boldsymbol{\phi}_i\|_2^2},\\ x[k] &:= \frac{\langle \boldsymbol{\phi}_{i[k]}, \boldsymbol{r}[k-1]\rangle}{\|\boldsymbol{\phi}_{i[k]}\|_2^2},\\ \boldsymbol{x}[k] &:= \boldsymbol{x}[k-1] + x[k]\boldsymbol{e}_{i[k]},\\ \boldsymbol{r}[k] &:= \boldsymbol{r}[k-1] - x[k]\boldsymbol{\phi}_{i[k]},\\ k &:= k+1. \end{aligned} \tag{5.32}$$

In this algorithm, eps is the *termination tolerance* that should be fixed beforehand.

Exercise 5.2. Prove that the following equality holds at the k-th step in the MP algorithm:

$$\|\boldsymbol{y}\|_2^2 = \sum_{j=1}^{k} \|x[j]\boldsymbol{\phi}_{i[j]}\|_2^2 + \|\boldsymbol{r}[k]\|_2^2. \tag{5.33}$$

Moreover, show that if $\boldsymbol{\phi}_1, \boldsymbol{\phi}_2, \ldots, \boldsymbol{\phi}_n$ are normalized, that is,

$$\|\boldsymbol{\phi}_i\|_2 = 1, \quad \forall i \in \{1, 2, \ldots, n\}, \tag{5.34}$$

then the following equality holds:

$$\|\boldsymbol{y}\|_2^2 = \sum_{j=1}^{k} |x[j]|^2 + \|\boldsymbol{r}[k]\|_2^2. \tag{5.35}$$

The following theorem gives the convergence property of the MP algorithm [89].

Theorem 5.2. Assume that dictionary $\{\phi_1, \phi_2, \ldots, \phi_n\}$ has m linearly independent vectors (i.e. $\operatorname{rank}(\Phi) = m$). Then there exists a constant $c \in (0, 1)$ such that

$$\|\boldsymbol{r}[k]\|_2^2 \leq c^k \|\boldsymbol{y}\|_2^2, \quad k = 0, 1, 2, \ldots. \tag{5.36}$$

From this theorem, it follows that the residual $\boldsymbol{r}[k]$ monotonically decreases and

$$\lim_{k \to \infty} \boldsymbol{r}[k] = \boldsymbol{0} \tag{5.37}$$

holds.

The convergence rate in (5.36) is *first-order* or *linear*, and the residual decreases exponentially, that is, $O(c^k)$. This rate is much faster than FISTA in (4.95) (p. 85) for the ℓ^1 regularization, which has $O(1/k^2)$ convergence.

5.2.2 Orthogonal matching pursuit (OMP)

We have seen that the residual $\boldsymbol{r}[k]$ by the matching pursuit (MP) algorithm (5.32) decreases very fast. However, in general, it does not always achieve $\boldsymbol{r}[k] = \boldsymbol{0}$ in a finite number of iterations, and the output vector $\boldsymbol{x}[k]$ for large k, or $\lim_{k \to \infty} \boldsymbol{x}[k]$ may not be sparse. This is because MP may choose an index $i[k]$ that was already chosen in previous steps. *Orthogonal Matching Pursuit* (OMP) is an algorithm to improve MP to achieve a finite number of iterations to obtain a sparse solution. This is done by removing an index from candidates if it was once chosen. Let us see the procedure of OMP precisely.

At the k-th step in MP, we choose the index by

$$i[k] = \underset{i \in \{1, \ldots, n\}}{\arg\max} \frac{\langle \boldsymbol{\phi}_i, \boldsymbol{r}[k-1] \rangle^2}{\|\boldsymbol{\phi}_i\|_2^2}, \quad \boldsymbol{r}[0] = \boldsymbol{y}, \quad k = 1, 2, \ldots \tag{5.38}$$

To memorize indices that were chosen in the previous steps, we define $\mathcal{S}_k$ as the set of indices chosen by the k-th step:

$$\mathcal{S}_k = \mathcal{S}_{k-1} \cup \{i[k]\}, \quad \mathcal{S}_0 = \emptyset, \quad k = 1, 2, \ldots \tag{5.39}$$

Also, let us define a linear subspace $\mathcal{C}_k$ of $\mathbb{R}^m$ spanned by vectors $\boldsymbol{\phi}_i$, $i \in \mathcal{S}_k$,

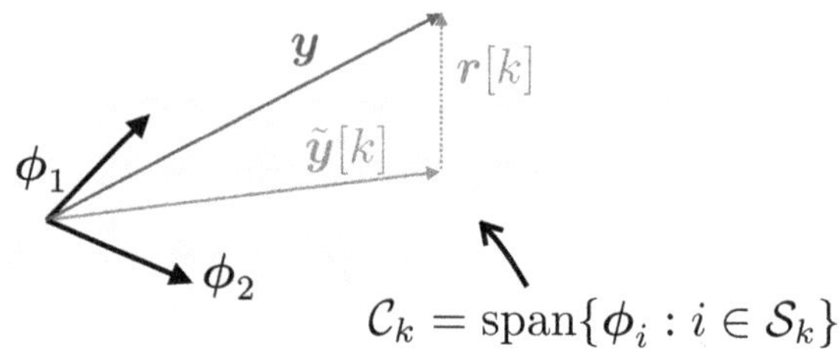

Figure 5.5: The k-th step of OMP: find the best approximation $\tilde{\boldsymbol{y}}[k]$ of $\boldsymbol{y}$ in the linear subspace $\mathcal{C}_k = \text{span}\{\boldsymbol{\phi}_i : i \in \mathcal{S}_k\}$. The residual vector $\boldsymbol{r}[k] = \boldsymbol{y} - \tilde{\boldsymbol{y}}[k]$ is orthogonal to $\mathcal{C}_k$.

that is,

$$\mathcal{C}_k \triangleq \text{span}\{\boldsymbol{\phi}_i : i \in \mathcal{S}_k\} = \left\{ \sum_{i \in \mathcal{S}_k} x_i \boldsymbol{\phi}_i : x_i \in \mathbb{R} \right\}. \tag{5.40}$$

OMP approximates the vector $\boldsymbol{y}$ at each step by a vector in $\mathcal{C}_k$, while MP approximates it by just one vector $\boldsymbol{\phi}_{i[k]}$. More precisely, OMP chooses a vector $\tilde{\boldsymbol{y}}[k]$ in $\mathcal{C}_k$ that has the minimum ℓ^2 distance from $\boldsymbol{y}$. This is obtained by the orthogonal projection of $\boldsymbol{y}$ onto $\mathcal{C}_k$:

$$\tilde{\boldsymbol{y}}[k] = \arg\min_{\boldsymbol{v} \in \mathcal{C}_k} \|\boldsymbol{v} - \boldsymbol{y}\|_2^2 = \Pi_{\mathcal{C}_k}(\boldsymbol{y}), \tag{5.41}$$

where $\Pi_{\mathcal{C}_k}$ is the projection operator onto $\mathcal{C}_k$. Figure 5.5 illustrates this projection at the k-th step.

Using the restriction notation,[1] we can characterize the condition $\boldsymbol{v} \in \mathcal{C}_k$ as

$$\boldsymbol{v} = \sum_{i \in \mathcal{S}_k} x_i \boldsymbol{\phi}_i = \Phi_{\mathcal{S}_k} \tilde{\boldsymbol{x}}, \tag{5.42}$$

for some $\tilde{\boldsymbol{x}} \in \mathbb{R}^k$. Note that $\#(\mathcal{S}_k) = k$ holds as explained later. Then, the projection in (5.41) is obtained by finding the coefficients of $\tilde{\boldsymbol{y}}[k]$ with respect to the basis functions $\boldsymbol{\phi}_i$, $i \in \mathcal{S}_k$ in $\mathcal{C}_k$. That is, we find

$$\tilde{\boldsymbol{x}}[k] = \arg\min_{\tilde{\boldsymbol{x}} \in \mathbb{R}^k} \frac{1}{2} \|\Phi_{\mathcal{S}_k} \tilde{\boldsymbol{x}} - \boldsymbol{y}\|_2^2. \tag{5.43}$$

This is the least squares solution[2]

$$\tilde{\boldsymbol{x}}[k] = \left(\Phi_{\mathcal{S}_k}^\top \Phi_{\mathcal{S}_k}\right)^{-1} \Phi_{\mathcal{S}_k}^\top \boldsymbol{y}. \tag{5.44}$$

Note that the matrix $\Phi_{\mathcal{S}_k}^\top \Phi_{\mathcal{S}_k}$ is always invertible (this will be explained later). Then $\tilde{\boldsymbol{y}}[k]$ is given by

$$\tilde{\boldsymbol{y}}[k] = \Phi_{\mathcal{S}_k} \tilde{\boldsymbol{x}}[k] = \Phi_{\mathcal{S}_k} \left(\Phi_{\mathcal{S}_k}^\top \Phi_{\mathcal{S}_k}\right)^{-1} \Phi_{\mathcal{S}_k}^\top \boldsymbol{y}. \tag{5.45}$$

[1]For the restriction notation, see (2.51), (2.52), and (2.53) in Chapter 2 (p. 28).
[2]For the least squares solution, see Section 3.1.2 in Chapter 3 and equation (3.24).

Define the coefficient vector $\boldsymbol{x}[k] \in \mathbb{R}^n$ with respect to $\boldsymbol{\phi}_i$, $i \in \{1, 2, \ldots, n\}$ by

$$(\boldsymbol{x}[k])_{\mathcal{S}_k} = \tilde{\boldsymbol{x}}[k], \quad (\boldsymbol{x}[k])_{\mathcal{S}_k^c} = \mathbf{0}, \tag{5.46}$$

where $\mathcal{S}_k^c$ is the complement of $\mathcal{S}_k$. Then we have

$$\tilde{\boldsymbol{y}}[k] = \Phi \boldsymbol{x}[k]. \tag{5.47}$$

The residual vector $\boldsymbol{r}[k] = \boldsymbol{y} - \tilde{\boldsymbol{y}}[k]$ is given by

$$\boldsymbol{r}[k] = \boldsymbol{y} - \tilde{\boldsymbol{y}}[k] = \{I - \Phi_{\mathcal{S}_k}(\Phi_{\mathcal{S}_k}^\top \Phi_{\mathcal{S}_k})^{-1}\Phi_{\mathcal{S}_k}^\top\}\boldsymbol{y}. \tag{5.48}$$

It is easily shown that the residual vector $\boldsymbol{r}[k]$ is orthogonal to the linear subspace $\mathcal{C}_k$ (see Figure 5.5), that is,

$$\langle \boldsymbol{v}, \boldsymbol{r}[k] \rangle = 0, \quad \forall \boldsymbol{v} \in \mathcal{C}_k. \tag{5.49}$$

This means that any vector $\boldsymbol{\phi}_i$ in $\mathcal{C}_k$ will never be chosen by the maximization at the next step:

$$i[k+1] = \underset{i \in \{1,2,\ldots,n\}}{\arg\max} \frac{\langle \boldsymbol{\phi}_i, \boldsymbol{r}[k] \rangle^2}{\|\boldsymbol{\phi}_i\|_2^2} = \underset{\substack{i \in \{1,2,\ldots,n\} \\ \boldsymbol{\phi}_i \notin \mathcal{C}_k}}{\arg\max} \frac{\langle \boldsymbol{\phi}_i, \boldsymbol{r}[k] \rangle^2}{\|\boldsymbol{\phi}_i\|_2^2}, \tag{5.50}$$

since $\langle \boldsymbol{\phi}_i, \boldsymbol{r}[k] \rangle = 0$ holds for any $\boldsymbol{\phi}_i \in \mathcal{C}_k$, from (5.49). Also, we see that $\boldsymbol{\phi}_i$, $i \in \mathcal{S}_k$ are always linearly independent since $\boldsymbol{\phi}_{i[k+1]} \notin \mathcal{C}_k$ holds for any k, and hence $\Phi_{\mathcal{S}_k}^\top \Phi_{\mathcal{S}_k}$ is invertible. The name *orthogonal* matching pursuit comes from this property of orthogonality.

We summarize the algorithm of OMP as follows.

OMP for ℓ^0 optimization (5.1)

Initialization: set $\boldsymbol{x}[0] = \mathbf{0}$, $\boldsymbol{r}[0] = \boldsymbol{y}$, $\mathcal{S}_0 = \emptyset$, and $k = 1$.
Iteration: while $\boldsymbol{r}[k] \neq \mathbf{0}$ do

$$\begin{aligned} i[k] &:= \underset{i \in \{1,\ldots,n\}}{\arg\max} \frac{\langle \boldsymbol{\phi}_i, \boldsymbol{r}[k-1] \rangle^2}{\|\boldsymbol{\phi}_i\|_2^2}, \\ \mathcal{S}_k &:= \mathcal{S}_{k-1} \cup \{i[k]\}, \\ \tilde{\boldsymbol{x}}[k] &:= (\Phi_{\mathcal{S}_k}^\top \Phi_{\mathcal{S}_k})^{-1}\Phi_{\mathcal{S}_k}^\top \boldsymbol{y}, \\ (\boldsymbol{x}[k])_{\mathcal{S}_k} &:= \tilde{\boldsymbol{x}}[k], \\ (\boldsymbol{x}[k])_{\mathcal{S}_k^c} &:= \mathbf{0}, \\ \boldsymbol{r}[k] &:= \boldsymbol{y} - \Phi_{\mathcal{S}_k}\tilde{\boldsymbol{x}}[k], \\ k &:= k+1. \end{aligned} \tag{5.51}$$

The following theorem shows that if there exists a sufficiently sparse solution of the equation $\Phi\boldsymbol{x} = \boldsymbol{y}$, then OMP gives the solution of the ℓ^0 optimization (5.1) in a finite number of iterations [43, Theorem 4.3]:

Theorem 5.3. Assume that $\Phi \in \mathbb{R}^{m\times n}$ is surjective, that is, $\mathrm{rank}(\Phi) = m$. Assume also that there exists a vector $\boldsymbol{x} \in \mathbb{R}^n$ such that $\Phi\boldsymbol{x} = \boldsymbol{y}$ and

$$\|\boldsymbol{x}\|_0 < \frac{1}{2}\left(1 + \frac{1}{\mu(\Phi)}\right), \tag{5.52}$$

where $\mu(\Phi)$ is the mutual coherence of matrix Φ. Then, this vector $\boldsymbol{x}$ is the unique solution of the ℓ^0 optimization (5.1), and OMP gives it in $k = \|\boldsymbol{x}\|_0$ steps.

We should note that at each step of OMP we need to compute the matrix inversion of $(\Phi_{\mathcal{S}_k}^\top \Phi_{\mathcal{S}_k})^{-1}\Phi_{\mathcal{S}_k}^\top \boldsymbol{y}$. If the number $k = \|\boldsymbol{x}\|_0$ in Theorem 5.3 is very large, then this inversion may impose a heavy computational burden.

5.3 Thresholding Algorithms

In this section, we consider the following optimization problems:

$$\underset{\boldsymbol{x}\in\mathbb{R}^n}{\text{minimize}} \quad \frac{1}{2}\|\Phi\boldsymbol{x} - \boldsymbol{y}\|_2^2 + \lambda\|\boldsymbol{x}\|_0, \tag{5.53}$$

$$\underset{\boldsymbol{x}\in\mathbb{R}^n}{\text{minimize}} \quad \frac{1}{2}\|\Phi\boldsymbol{x} - \boldsymbol{y}\|_2^2 \ \text{ subject to } \ \|\boldsymbol{x}\|_0 \leq s. \tag{5.54}$$

The first problem (5.53) is called the ℓ^0 *regularization*, and the second problem (5.54) is called the *s-sparse approximation*. Note that these optimization problems are non-convex and combinatorial. For these problems, we introduce efficient greedy algorithms by borrowing the idea of the proximal gradient algorithm introduced in Chapter 4.

5.3.1 Iterative hard-thresholding algorithm (IHT)

Let us consider the following optimization problem:

$$\underset{\boldsymbol{x}\in\mathbb{R}^n}{\text{minimize}}\, f_1(\boldsymbol{x}) + f_2(\boldsymbol{x}), \tag{5.55}$$

where f_1 is a differentiable and convex function satisfying $\mathrm{dom}(f_1) = \mathbb{R}^n$, and f_2 is a proper, closed, and convex function. The proximal gradient

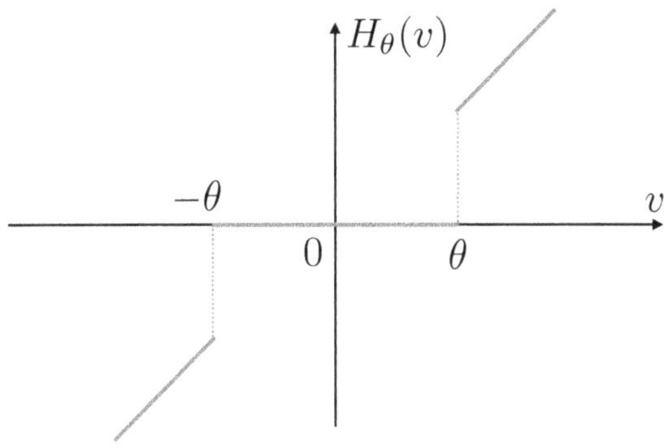

Figure 5.6: Hard-thresholding operator $H_\theta(v)$

algorithm for this is given by[3]

$$\boldsymbol{x}[k+1] = \mathrm{prox}_{\gamma f_2}(\boldsymbol{x}[k] - \gamma \nabla f_1(\boldsymbol{x}[k])). \tag{5.56}$$

For the ℓ^0 regularization of (5.53), we have

$$f_1(\boldsymbol{x}) \triangleq \frac{1}{2}\|\Phi \boldsymbol{x} - \boldsymbol{y}\|_2^2, \quad f_2(\boldsymbol{x}) \triangleq \lambda \|\boldsymbol{x}\|_0. \tag{5.57}$$

Although the function f_2 is not convex in this case, we naively apply the proximal gradient algorithm (5.56). Now, the proximal operator of $f_2(\boldsymbol{x}) = \lambda\|\boldsymbol{x}\|_0$ has a closed form, called the *hard-thresholding operator* (see Figure 5.6), defined by

$$[H_\theta(\boldsymbol{v})]_i \triangleq \begin{cases} v_i, & |v_i| \geq \theta, \\ 0, & |v_i| < \theta, \quad i = 1, 2, \ldots, n, \end{cases} \tag{5.58}$$

with $\theta = \sqrt{2\gamma\lambda}$. That is,

$$\mathrm{prox}_{\gamma f_2}(\boldsymbol{v}) = H_{\sqrt{2\gamma\lambda}}(\boldsymbol{v}). \tag{5.59}$$

See Exercise 4.12 (p. 76) for details. As shown in Figure 5.6, the hard-thresholding operator rounds small elements ($|v_i| < \theta$) to 0. Figure 5.7 illustrates this operation. By using this operator, the proximal gradient algorithm for the ℓ^0 regularization (5.53) is given as follows:

[3]See Section 4.4 (p. 81).

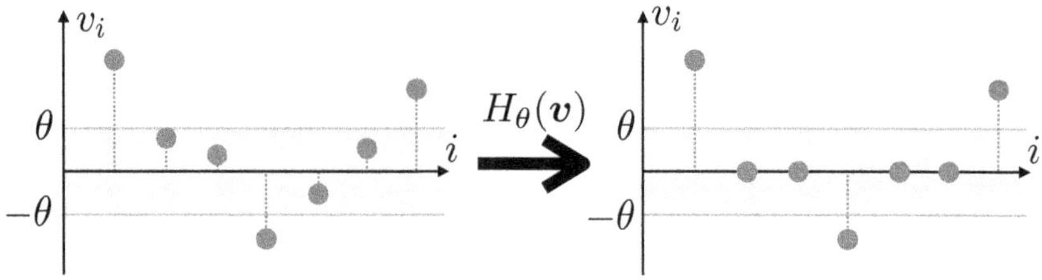

Figure 5.7: Hard-thresholding operator $H_\theta(\boldsymbol{v})$ rounds small elements ($|v_i| < \theta$) to 0, where $\theta = \sqrt{2\gamma\lambda}$.

IHT for ℓ^0 regularization (5.53)

Initialization: give an initial vector $\boldsymbol{x}[0]$ and positive number $\gamma > 0$.
Iteration: for $k = 0, 1, 2, \ldots$ do

$$\boldsymbol{x}[k+1] = H_{\sqrt{2\gamma\lambda}}(\boldsymbol{x}[k] - \gamma\Phi^\top(\Phi\boldsymbol{x}[k] - \boldsymbol{y})). \tag{5.60}$$

This algorithm is called the *iterative hard-thresholding algorithm* (IHT).

For the convergence of the iterative hard-thresholding algorithm (5.60), the following theorem is proved in [11]:

Theorem 5.4. Assume that

$$\gamma < \frac{1}{\|\Phi\|^2} \tag{5.61}$$

holds where $\|\Phi\|$ is the maximum singular value of Φ. Then the sequence $\{\boldsymbol{x}[0], \boldsymbol{x}[1], \boldsymbol{x}[2], \ldots\}$ generated by the iterative hard-thresholding algorithm (5.60) converges to a local minimizer of the ℓ^0 regularization (5.53). Moreover, the convergence is first-order, that is, there exists a constant $c \in (0, 1)$ such that

$$\|\boldsymbol{x}[k+1] - \boldsymbol{x}^*\|_2 \leq c\|\boldsymbol{x}[k] - \boldsymbol{x}^*\|_2, \quad k = 0, 1, 2, \ldots, \tag{5.62}$$

where $\boldsymbol{x}^*$ is a local minimizer.

The condition in (5.61) is very similar to the condition in (4.94) (p. 85) for the proximal gradient algorithm for ℓ^1 regularization. The convergence rate $O(c^k)$ of IHT is much faster than that of ISTA, $O(1/k)$, or FISTA, $O(1/k^2)$. Note also that IHT may only converge to a local minimum, which may not necessarily be identical to the global minimum.

5.3.2 Iterative s-sparse algorithm

Here we consider the s-sparse approximation (5.54). By using the indicator function (4.37) in Chapter 4 (p. 72), we rewrite the constrained problem of s-sparse approximation (5.54) as an unconstrained optimization problem. Let Σ_s denote the set of s-sparse vectors in $\mathbb{R}^n$, that is,

$$\Sigma_s \triangleq \{\boldsymbol{x} \in \mathbb{R}^n : \|\boldsymbol{x}\|_0 \leq s\}. \tag{5.63}$$

Exercise 5.3. Show that Σ_s is a non-convex set.

The indicator function I_{Σ_s} for the set Σ_s is given by

$$I_{\Sigma_s}(\boldsymbol{x}) = \begin{cases} 0, & \|\boldsymbol{x}\|_0 \leq s, \\ \infty, & \|\boldsymbol{x}\|_0 > s. \end{cases} \tag{5.64}$$

By using this, the s-sparse approximation (5.54) is equivalently described by

$$\underset{\boldsymbol{x} \in \mathbb{R}^n}{\text{minimize}} \quad \frac{1}{2}\|\Phi\boldsymbol{x} - \boldsymbol{y}\|_2^2 + I_{\Sigma_s}(\boldsymbol{x}). \tag{5.65}$$

Note that since Σ_s is non-convex (see Exercise 5.3), the indicator function I_{Σ_s} is not a convex function. Anyhow, let us apply this to the proximal gradient algorithm (5.56). To do this, we should compute the proximal operator of the indicator function I_{Σ_s}, which is equal to the projection onto the set Σ_s. The projection is actually obtained by

$$\Pi_{\Sigma_s}(\boldsymbol{v}) = \underset{\boldsymbol{x} \in \Sigma_s}{\arg\min} \|\boldsymbol{x} - \boldsymbol{v}\|_2 = \mathcal{H}_s(\boldsymbol{v}), \tag{5.66}$$

where $\mathcal{H}_s(\boldsymbol{v})$ is the *s-sparse operator* that sets all but the s largest (in magnitude) elements of $\boldsymbol{v}$ to 0. Figure 5.8 illustrates this operation. Note that the projection is in general not unique. If s largest elements are not uniquely determined, then they can be chosen either randomly or based on a fixed ordering rule.

Exercise 5.4. Prove that the equation (5.66) holds.

Let $\gamma_s(\boldsymbol{v})$ denote the s-th largest element of vector $\boldsymbol{v} \in \mathbb{R}^n$. Then the s-sparse operator $\mathcal{H}_s(\boldsymbol{v})$ can be represented by using the hard-thresholding operator (5.58) as

$$\mathcal{H}_s(\boldsymbol{v}) = H_{\gamma_s(\boldsymbol{v})}(\boldsymbol{v}). \tag{5.67}$$

By using the s-sparse operator (5.66) as the proximal operator of the indicator function I_{Σ_s}, we obtain the proximal gradient algorithm (5.56) for the s-sparse approximation (5.65).

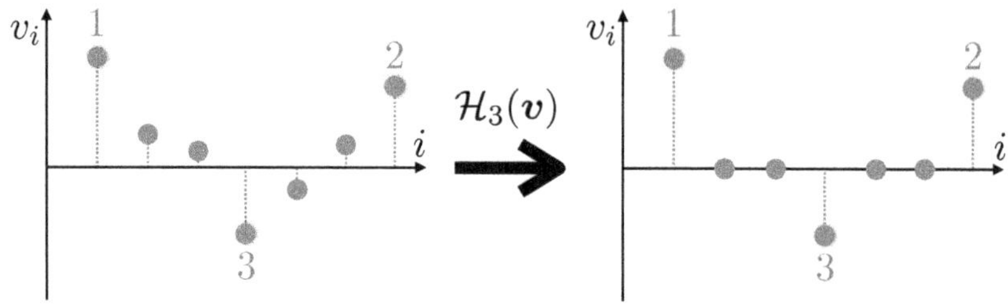

Figure 5.8: s-sparse operator $\mathcal{H}_s(\boldsymbol{v})$ with $s = 3$: the 3 largest elements in magnitude are unchanged and the other elements are set to 0. The numbers 1, 2, 3 indicates the rank of the absolute values of the elements.

Iterative s-sparse algorithm for s-sparse approximation (5.54)

Initialization: give an initial vector $\boldsymbol{x}[0]$ and a positive number $\gamma > 0$.
Iteration: for $k = 0, 1, 2, \ldots$ do

$$\boldsymbol{x}[k+1] = \mathcal{H}_s\big(\boldsymbol{x}[k] - \gamma\Phi^\top(\Phi\boldsymbol{x}[k] - \boldsymbol{y})\big). \tag{5.68}$$

We call this algorithm the *iterative s-sparse algorithm.*

For the iterative s-sparse algorithm, we have the following convergence theorem [11].

Theorem 5.5. Assume that the matrix $\Phi \in \mathbb{R}^{m\times n}$ is surjective, that is, $\operatorname{rank}(\Phi) = m$, and the column vectors ϕ_i, $i = 1, 2, \ldots, n$, are non-zero, that is,

$$\|\phi_i\|_2 > 0, \quad \forall i \in \{1, 2, \ldots, n\}. \tag{5.69}$$

Assume also that the constant $\gamma > 0$ satisfies

$$\gamma < \frac{1}{\|\Phi\|^2}. \tag{5.70}$$

Then the sequence $\{\boldsymbol{x}[0], \boldsymbol{x}[1], \boldsymbol{x}[2], \ldots\}$ generated by the s-sparse algorithm (5.68) converges to a local minimizer of the s-sparse approximation problem (5.54). Moreover, the convergence is first-order, that is, there exists a constant $c \in (0, 1)$ such that

$$\|\boldsymbol{x}[k+1] - \boldsymbol{x}^*\|_2 \le c\|\boldsymbol{x}[k] - \boldsymbol{x}^*\|_2, \quad k = 0, 1, 2, \ldots, \tag{5.71}$$

where $\boldsymbol{x}^*$ is a local minimizer.

5.3.3 Compressive sampling matching pursuit (CoSaMP)

For the s-sparse approximation (5.54), we can extend the algorithm of OMP in Section 5.2.2 with the s-sparse operator $\mathcal{H}_s$. This algorithm is called the *compressive sampling matching pursuit* (CoSaMP).

In the OMP algorithm (5.51), we choose one index $i[k]$ as

$$i[k] = \underset{i \in \{1,\dots,n\}}{\arg\max} \frac{\langle \boldsymbol{\phi}_i, \boldsymbol{r}[k-1] \rangle^2}{\|\boldsymbol{\phi}_i\|_2^2}. \tag{5.72}$$

Alternatively, CoSaMP chooses $2s$ largest values of

$$\frac{\langle \boldsymbol{\phi}_i, \boldsymbol{r}[k-1] \rangle^2}{\|\boldsymbol{\phi}_i\|_2^2} = \left\langle \frac{\boldsymbol{\phi}_i}{\|\boldsymbol{\phi}_i\|_2}, \boldsymbol{r}[k-1] \right\rangle^2, \tag{5.73}$$

and includes these $2s$ indices in the index set $\mathcal{S}_k$, that is,

$$\mathcal{S}_k = \mathcal{S}_{k-1} \cup \operatorname{supp}\left\{ \mathcal{H}_{2s}\left(\left\langle \frac{\boldsymbol{\phi}_i}{\|\boldsymbol{\phi}_i\|_2}, \boldsymbol{r}[k-1] \right\rangle^2 \right) \right\}. \tag{5.74}$$

As in OMP, we then find the projection of $\boldsymbol{y}$ onto the linear subspace $\mathcal{C}_k = \{\boldsymbol{\phi}_i : i \in \mathcal{S}_k\}$. That is,

$$\tilde{\boldsymbol{x}}[k] = \left(\Phi_{\mathcal{S}_k}^\top \Phi_{\mathcal{S}_k}\right)^{-1} \Phi_{\mathcal{S}_k}^\top \boldsymbol{y}. \tag{5.75}$$

From this, we define an n-dimensional coefficient vector $\boldsymbol{z}[k]$ as

$$(\boldsymbol{z}[k])_i \triangleq \begin{cases} (\tilde{\boldsymbol{x}}[k])_i, & i \in \mathcal{S}_k, \\ 0, & i \notin \mathcal{S}_k. \end{cases} \tag{5.76}$$

Note that the number of nonzero coefficients in $\boldsymbol{z}[k]$ is larger than $2s$. We then *prune* $\boldsymbol{z}[k]$ to an s-sparse vector $\boldsymbol{x}[k]$ as

$$\boldsymbol{x}[k] = \mathcal{H}_s(\boldsymbol{z}[k]). \tag{5.77}$$

Also, we update the index set $\mathcal{S}_k$ to

$$\mathcal{S}_k = \operatorname{supp}(\boldsymbol{x}[k]). \tag{5.78}$$

Finally, we obtain the CoSaMP algorithm to solve the s-sparse approximation (5.54).

CoSaMP algorithm for s-sparse approximation (5.54)

Initialization: set $\boldsymbol{x}[0] = \mathbf{0}$, $\boldsymbol{r}[0] = \boldsymbol{y}$, and $\mathcal{S}_0 = \emptyset$.
Iteration: for $k = 1, 2, \dots$ do

$$\begin{aligned}
\mathcal{I}[k] &:= \operatorname{supp}\left\{\mathcal{H}_{2s}\left(\left\langle \frac{\boldsymbol{\phi}_i}{\|\boldsymbol{\phi}_i\|_2}, \boldsymbol{r}[k-1]\right\rangle^2\right)\right\},\\
\mathcal{S}_k &:= \mathcal{S}_{k-1} \cup \mathcal{I}[k],\\
\tilde{\boldsymbol{x}}[k] &:= \left(\Phi_{\mathcal{S}_k}^\top \Phi_{\mathcal{S}_k}\right)^{-1} \Phi_{\mathcal{S}_k}^\top \boldsymbol{y},\\
(\boldsymbol{z}[k])_{\mathcal{S}_k} &:= \tilde{\boldsymbol{x}}[k],\\
(\boldsymbol{z}[k])_{\mathcal{S}_k^c} &:= \mathbf{0},\\
\boldsymbol{x}[k] &:= \mathcal{H}_s\left(\boldsymbol{z}[k]\right),\\
\mathcal{S}_k &:= \operatorname{supp}\{\boldsymbol{x}[k]\},\\
\boldsymbol{r}[k] &:= \boldsymbol{y} - \Phi_{\mathcal{S}_k} \tilde{\boldsymbol{x}}[k].
\end{aligned} \tag{5.79}$$

For the convergence of the CoSaMP algorithm, see the original paper [118].

5.4 Numerical Example

Here we solve sparse optimization numerically by using greedy algorithms studied in this chapter. Let us consider the problem of curve fitting studied in Section 3.3 (p. 54) with the sparse polynomial $y = -t^{80} + t$. As in Section 3.3, the data points are given by

$$t_i = 0.1(i-1), \quad i = 1, 2, \dots, 11, \tag{5.80}$$

from which we reconstruct the 80-th order polynomial. Here we consider the following 6 algorithms:

1. ℓ^1 optimization considered in Section 3.3 (p. 54)
2. matching pursuit (MP)
3. orthogonal matching pursuit (OMP)
4. iterative hard-thresholding (IHT)
5. iterative s-sparse algorithm (ISS)
6. compressive sampling matching pursuit (CoSaMP)

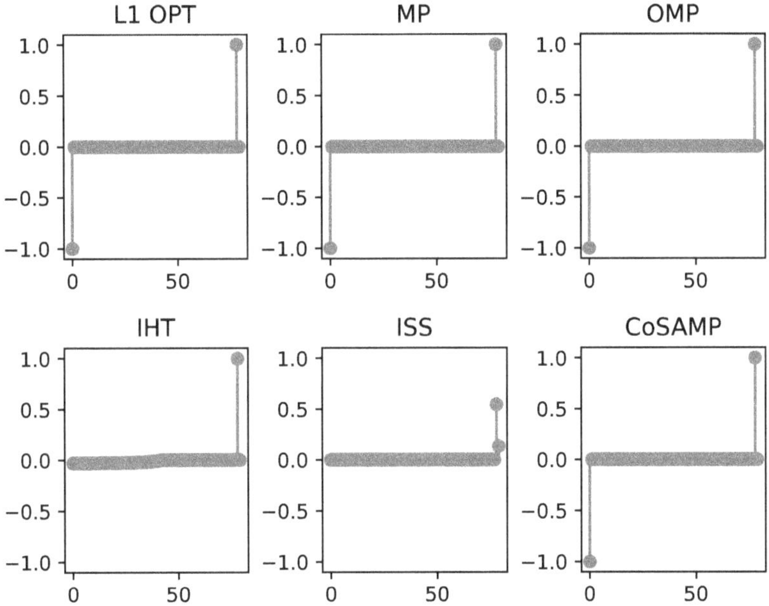

Figure 5.9: Estimation of sparse coefficients

The matrix $\Phi \in \mathbb{R}^{11\times 81}$ is the Vandermonde matrix defined by (3.16), which satisfies

$$0.012 < \frac{1}{\|\Phi\|^2} < 0.013. \tag{5.81}$$

We choose the parameter γ for IHT and ISS as

$$\gamma = 0.01 < 0.012 < \frac{1}{\|\Phi\|^2}. \tag{5.82}$$

From Theorems 5.4 and 5.5, the condition (5.82) guarantees convergence to local minimizers for IHT and ISS. We also choose λ in the ℓ^0 regularization problem as $\lambda = 0.001$.

Figure 5.9 shows the coefficients obtained by the algorithms. The coefficients are ordered from the highest degree to the lowest degree. We see that the ℓ^1 optimization, MP, OMP, and CoSaMP give exact coefficients, while IHT and ISS show incorrect reconstruction. To see this more precisely, we check the estimation error $\boldsymbol{r} = \boldsymbol{y} - \Phi\tilde{\boldsymbol{x}}$, where $\tilde{\boldsymbol{x}}$ is the obtained vector when the algorithm stops. Table 5.1 shows the error with the number of iterations required to achieve the error.

Table 5.1: Estimation error $\|\boldsymbol{y} - \Phi\tilde{\boldsymbol{x}}\|_2$ and number of iterations. IHT and ISS reached the maximum number 10^5 of iterations.

Methods	ℓ^1 OPT	MP	OMP	IHT	ISS	CoSaMP
Error	2.8×10^{-12}	9.1×10^{-6}	2.5×10^{-16}	0.0017	0.83	4.1×10^{-16}
Iterations	9	18	2	10^5	10^5	3

All but ℓ^1 optimization stop the iteration when the error $\|\boldsymbol{r}[k]\|_2$ is less than 10^{-5} or the number of iterations is larger than 10^5.

IHT and ISS attained the maximum number of iterations 10^5, and their errors are much larger than those of the other methods. This is because they were trapped into local minimizers. The other methods show fast convergence, among which OMP (2 iterations) and CoSaMP (3 iterations) especially present surprising results. In view of the error and the number of iterations, OMP is the best method in this case. It should be noted that greedy algorithms do not necessarily give a global solution. OMP is the best in this case but in other cases, another method may be the best. This depends on the problem and data, and we should adopt trial and error to seek the best algorithm.

5.5 Further Readings

Basics of greedy algorithms can be found in [30], [75]. For the characterization of ℓ^0 optimality by using the mutual coherence and the *restricted isometry property* (RIP), you can refer to [43], [45].

The matching pursuit (MP) was first proposed in [89], while the orthogonal matching pursuit (OMP) was introduced in [33], [124]. For the iterative hard-thresholding algorithm and the iterative s-sparse algorithm, see the paper [11]. The compressive sampling matching pursuit (CoSaMP) was proposed in [118].

5.6 Python Programs

The following program defines the algorithms studied in this chapter.

```
import numpy as np

# Matching Pursuit (MP) algorithm
```

```
def MP(y, Phi, EPS=1e-5, MAX_ITER=10000):
    m, n = Phi.shape
    x = np.zeros(n)
    r = y
    k = 0
    Phi_norm = np.diag(Phi.T @ Phi)
    while np.linalg.norm(r) > EPS and k < MAX_ITER
        :
        p = Phi.T @ r
        v = p / np.sqrt(Phi_norm)
        ik = np.argmax(np.abs(v))
        v2 = p / Phi_norm
        z = v2[ik]
        x[ik] += z
        r -= z * Phi[:, ik]
        k += 1
    nitr = k
    return x, nitr

# Orthogonal Matching Pursuit (OMP) algorithm
def OMP(y, Phi, EPS=1e-5, MAX_ITER=10000):
    m, n = Phi.shape
    x = np.zeros(n)
    r = y
    k = 0
    S = []
    Phi_norm = np.diag(Phi.T @ Phi)
    while (np.linalg.norm(r) > EPS) and (k <
        MAX_ITER):
        p = Phi.T @ r
        v = p / np.sqrt(Phi_norm)
        ik = np.argmax(np.abs(v))
        if ik not in S:
            S.append(ik)
        Phi_S = Phi[:, S]
        x_S = np.linalg.lstsq(Phi_S, y, rcond=None
            )[0]
        x = np.zeros(n)
```

```
        x[S] = x_S
        r = y - Phi @ x
        k += 1

    nitr = k
    return x, nitr

# Hard thresholding operator
def hard_thresholding(lambda_val, v):
    hv = np.where(np.abs(v) <= lambda_val, 0, v)
    return hv

# Support function
def supp(x):
    return np.nonzero(np.abs(x) > 0)[0]

# Iterative Hard Thresholding (IHT) algorithm
def IHT(y, Phi, lambda_val=1, gamma=1, EPS=1e-5,
        MAX_ITER=10000):
    m, n = Phi.shape
    x = np.zeros(n)
    r = y
    k = 0
    while np.linalg.norm(r) > EPS and k < MAX_ITER
        :
        p = x + gamma * (Phi.T @ r)
        x = hard_thresholding(np.sqrt(2 *
            lambda_val * gamma), p)
        S = supp(x)
        r = y - Phi[:, S] @ x[S]
        k += 1
    nitr = k
    return x, nitr

# s-sparse operator
def s_sparse_operator(A, s):
    x = A.flatten()
    y = np.zeros_like(x)
```

```
    indx = np.argsort(-np.abs(x))[:s]
    y[indx] = x[indx]
    return y.reshape(A.shape)

# Iterative s-sparse algorithm
def iterative_s_sparse(y, Phi, s, gamma=1, EPS=1e
   -5, MAX_ITER=10000):
    m, n = Phi.shape
    x = np.zeros(n)
    r = y
    k = 0
    while np.linalg.norm(r) > EPS and k < MAX_ITER
        :
        p = x + gamma * (Phi.T @ r)
        x = s_sparse_operator(p, s)
        S = supp(x)
        r = y - Phi[:, S] @ x[S]
        k += 1
    nitr = k
    return x, nitr

# CoSaMP algorithm
def CoSaMP(y, Phi, s, EPS=1e-5, MAX_ITER=10000):
    m, n = Phi.shape
    x = np.zeros(n)
    r = y
    k = 0
    S = []
    Lambda = []
    Phi_norm = np.diag(Phi.T @ Phi)
    while np.linalg.norm(r) > EPS and k < MAX_ITER
        :
        p = s_sparse_operator((Phi.T @ r) / np.
           sqrt(Phi_norm), 2 * s)
        Ik = supp(p)
        S = np.union1d(Lambda, Ik)
        Phi_S = Phi[:, S.astype(int)]
        z = np.zeros(n)
```

```
            z[S.astype(int)] = np.linalg.pinv(Phi_S) @
                y
            x = s_sparse_operator(z, s)
            Lambda = supp(x)
            r = y - Phi_S @ z[S.astype(int)]
            k += 1
        nitr = k
        return x, nitr
```

The following is the program to obtain the result by OMP. Other algorithms can be tested by changing the 29th line to one of the algorithms defined above.

```
1  import numpy as np
2  import matplotlib.pyplot as plt
3  import cvxpy as cp
4
5  # Polynomial coefficients
6  x_orig = np.zeros(80)
7  x_orig[0] = -1
8  x_orig[78] = 1
9
10 # Data
11 t = np.arange(0, 1.1, 0.1)
12 y = np.polyval(x_orig, t)
13
14 # Data size
15 N = len(t)
16 M = N - 1
17
18 # Order of polynomial
19 M_1 = len(x_orig) - 1
20
21 # Vandermonde matrix
22 Phi = np.vander(t, N=M_1+1)
23
24 # Parameters for iteration
25 EPS = 1e-5  # if the residue < EPS then the
       iteration will stop
```

```
MAX_ITER = 100000  # maximum number of iterations

# OMP
x_omp, nitr_omp = OMP(y, Phi,EPS,10)
# Result
fig = plt.figure()
ax1 = fig.add_subplot(1, 2, 1)
ax1.stem(x_orig)
ax1.set_title("Original")
plt.ylim(-1.1,1.1)
ax2 = fig.add_subplot(1, 2, 2)
ax2.stem(x_omp)
ax2.set_title("OMP")
plt.ylim(-1.1,1.1)
```

Chapter 6

Distributed Optimization

Distributed optimization focuses on solving an optimization problem by dividing it into smaller sub-problems, which are then solved by a lot of small processors called *agents* working collaboratively over a *network*. This approach can leverage the computational power of multiple processors, which is suitable for handling large-scale optimization problems that are difficult to solve on a single processor.

Key ideas of Chapter 6

- Algebraic graph theory is a fundamental tool for the analysis and design of distributed optimization algorithms.
- Distributed optimization algorithms solve an optimization problem by multiple computer agents, each of which solves a local subproblem while exchanging information with its neighbors over a network.
- The distributed gradient descent algorithm is derived from the combination of gradient descent and average consensus control.
- The ADMM algorithm is extended to a distributed algorithm by employing the consensus set. This distributed implementation uses a special agent called the central collector.

6.1 Network Model and Algebraic Graph Theory

To study distributed optimization, we need the knowledge of *networks*. A network is mathematically modeled by a *graph*, which consists of a finite

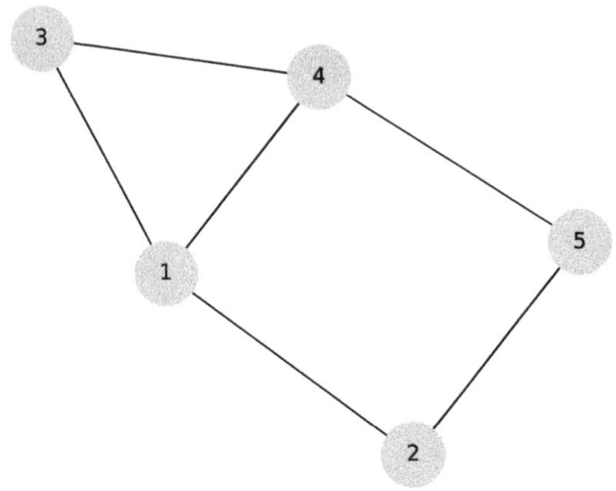

Figure 6.1: An undirected graph $G = (\mathcal{V}, \mathcal{E})$.

number of *nodes* (or *vertices*) with *edges* that connect pairs of nodes.

Let $\mathcal{V} = \{1, 2, \ldots, n\}$ be the set of n nodes (or vertices) in a network, and let $\mathcal{E} \subset \mathcal{V} \times \mathcal{V}$ be the set of edges in a network. The graph G with the node set $\mathcal{V}$ and the edge set $\mathcal{E}$ is defined as the pair $(\mathcal{V}, \mathcal{E})$. There are two types of graphs: *undirected graphs* and *directed graphs* (or *digraphs*). For a directed graph $G = (\mathcal{V}, \mathcal{E})$, $(i, j) \in \mathcal{E}$ $(i, j \in \mathcal{V})$ does not always imply $(j, i) \in \mathcal{E}$. For an undirected graph, on the other hand, $(i, j) \in \mathcal{E}$ holds if and only if $(j, i) \in \mathcal{E}$ holds. Therefore, for an undirected graph, we only include (i, j) in $\mathcal{E}$ when $i < j$ and exclude (j, i) from $\mathcal{E}$ for simplicity. In this chapter, we only consider undirected graphs and adopt this simple notation. Also, we assume

- there is no self-connection, which is an edge starting and ending at the same node,
- and there is at most one edge between any two vertices (no multiple edges).

Such a graph is called a *simple graph*.

Example 6.1. Figure 6.1 shows an undirected graph $G = (\mathcal{V}, \mathcal{E})$ with $\mathcal{V} = \{1, 2, 3, 4, 5\}$ and $\mathcal{E} = \{(1, 2), (1, 3), (1, 4), (2, 5), (3, 4), (4, 5)\}$. □

Let us consider an undirected graph $G = (\mathcal{V}, \mathcal{E})$. A *neighbor*, or an *adjacent node*, of $i \in \mathcal{V}$ is a node connected to i by an edge in $\mathcal{E}$. We denote by $\mathcal{N}_i$ the set of all neighbors of $i \in \mathcal{V}$. The number of nodes adjacent to $i \in \mathcal{V}$, or the number of entries in $\mathcal{N}_i$, is called the *degree* of node i, which is denoted by d_i. The *maximum degree*, denoted by Δ, is the maximum degree among all nodes, that is,

$$\Delta \triangleq \max\{d_i : i \in \mathcal{V}\}. \tag{6.1}$$

To *numerically* analyze a graph, particularly for large-scale graphs, we employ techniques from linear algebra. The method of graph theory based on linear algebra is called the *algebraic graph theory*. For this, we define some matrices related to a graph.

The *adjacency matrix* $A = [a_{ij}]$ of a graph $G = (\mathcal{V}, \mathcal{E})$ with $\mathcal{V} = \{1, 2, \ldots, n\}$ is defined by

$$a_{ij} \triangleq \begin{cases} 1, & \text{if } (i,j) \in \mathcal{E}, \\ 0, & \text{otherwise.} \end{cases} \tag{6.2}$$

We note that A is a symmetric binary-valued $n \times n$ matrix since G is undirected. Also, we have $a_{ii} = 0$ for any $i \in \mathcal{V}$ since G is simple.

The *degree matrix* D is a diagonal matrix defined by

$$D \triangleq \operatorname{diag}(d_1, d_2, \ldots, d_n) = \begin{bmatrix} d_1 & 0 & \ldots & 0 \\ 0 & d_2 & \ddots & \vdots \\ \vdots & \ddots & \ddots & 0 \\ 0 & \ldots & 0 & d_n \end{bmatrix}. \tag{6.3}$$

With the adjacency matrix A and the degree matrix D, the *graph Laplacian* is defined by

$$L \triangleq D - A. \tag{6.4}$$

We note that we have

$$d_i = \sum_{j=1}^{n} a_{ij}, \tag{6.5}$$

and hence the graph Laplacian is represented as

$$L = \begin{bmatrix} \sum_{j=1}^{n} a_{1j} & -a_{12} & \cdots & -a_{1n} \\ -a_{21} & \sum_{j=1}^{n} a_{2j} & \ddots & \vdots \\ \vdots & \ddots & \ddots & -a_{n-1,n} \\ -a_{n1} & \cdots & -a_{n,n-1} & \sum_{j=1}^{n} a_{nj} \end{bmatrix}. \tag{6.6}$$

By using the graph Laplacian, we can check the connectivity of a graph. Two nodes $i, j \in \mathcal{V}$ are said to be *connected* if there exists a path between i and j. A graph is said to be *connected* if any nodes $i, j \in \mathcal{V}$ are connected to each other. We have the following theorem [98, Corollary 2.1]:

Theorem 6.1. Let G be an undirected simple graph. G is connected if and only if the eigenvalue 0 of the graph Laplacian of G is simple.

We note that the graph Laplacian L has at least one zero eigenvalue, since we have

$$L \begin{bmatrix} 1 \\ 1 \\ \vdots \\ 1 \end{bmatrix} = \mathbf{0}. \tag{6.7}$$

Example 6.2. Let us consider the graph in Fig.6.1. The adjacency matrix A and the degree matrix D are given by

$$A = \begin{bmatrix} 0 & 1 & 1 & 1 & 0 \\ 1 & 0 & 0 & 0 & 1 \\ 1 & 0 & 0 & 1 & 0 \\ 1 & 0 & 1 & 0 & 1 \\ 0 & 1 & 0 & 1 & 0 \end{bmatrix}, \quad D = \begin{bmatrix} 3 & 0 & 0 & 0 & 0 \\ 0 & 2 & 0 & 0 & 0 \\ 0 & 0 & 2 & 0 & 0 \\ 0 & 0 & 0 & 3 & 0 \\ 0 & 0 & 0 & 0 & 2 \end{bmatrix}. \tag{6.8}$$

The graph Laplacian $L = A - D$ is then given by

$$L = \begin{bmatrix} 3 & -1 & -1 & -1 & 0 \\ -1 & 2 & 0 & 0 & -1 \\ -1 & 0 & 2 & -1 & 0 \\ -1 & 0 & -1 & 3 & -1 \\ 0 & -1 & 0 & -1 & 2 \end{bmatrix}. \tag{6.9}$$

The eigenvalues of L are numerically computed as $\{0, 4.62, 3.62, 2.38, 1.38\}$, and hence the zero eigenvalue is simple. The graph shown in Fig. 6.1 is obviously connected, but this can be proved by numerical computation, thanks to Theorem 6.1. □

6.2 Consensus Algorithm

Let us consider an undirected graph $G = (\mathcal{V}, \mathcal{E})$ with $\mathcal{V} = \{1, 2, \ldots, n\}$. Each node $i \in \mathcal{V}$ represents a small computer called an *agent*. Agent i can process information from its neighbors $j \in \mathcal{N}_i$, and update its *state* x_i according to

$$x_i[k+1] = x_i[k] + u_i[k], \quad k = 0, 1, 2, \ldots, \tag{6.10}$$

where $u_i[k]$ is determined by the current state $x_i[k]$ of agent i and the state $x_j[k]$ of its neighbors $j \in \mathcal{N}_i$ at discrete time $k \in \{0, 1, 2, \ldots\}$. The variable $u_i[k]$ is called the *local control*, for which we consider the following linear control:

$$u_i[k] = \epsilon \sum_{j \in \mathcal{N}_i} (x_j[k] - x_i[k]). \tag{6.11}$$

The parameter $\epsilon > 0$ is called the *control gain*, or the *step size*. We can rewrite (6.11) as

$$u_i[k] = \epsilon d_i \times \frac{1}{d_i} \sum_{j \in \mathcal{N}_i} (x_j[k] - x_i[k]), \tag{6.12}$$

where d_i is the degree of node i, or the number of nodes in $\mathcal{N}_i$, and $u_i[k]$ is proportional to the *average* of the difference $x_j[k] - x_i[k]$, $j \in \mathcal{N}_i$. This property is important to analyze the limiting value of $x_i[k]$ (see Theorem 6.2 below). Inserting (6.11) into (6.10) gives

$$\boldsymbol{x}[k+1] = P_\epsilon \boldsymbol{x}[k], \tag{6.13}$$

where

$$\boldsymbol{x}[k] \triangleq \begin{bmatrix} x_1[k] \\ x_2[k] \\ \vdots \\ x_n[k] \end{bmatrix}, \quad P_\epsilon \triangleq I - \epsilon L, \tag{6.14}$$

and L is the graph Laplacian defined in (6.6). The matrix P_ϵ is called the *Perron matrix*.

In this section, we discuss the *consensus* of the network system (6.13).

Definition 6.1. The network system (6.13) is said to *achieve consensus* if

$$\lim_{k \to \infty} |x_i[k] - x_j[k]| = 0, \tag{6.15}$$

for any $i, j \in \mathcal{V}$.

The following theorem shows the condition for consensus.

Theorem 6.2. Let Δ be the maximum degree of the undirected graph $G = (\mathcal{V}, \mathcal{E})$. Suppose G is connected and $\epsilon\Delta < 1$ holds. Then the network system (6.13) achieves consensus, and the consensus value is given by the average of the initial states, that is,

$$\lim_{k \to \infty} x_i[k] = \overline{\boldsymbol{x}[0]} = \frac{1}{n} \sum_{j=1}^{n} x_j[0], \tag{6.16}$$

holds for any $i \in \mathcal{V}$.

Since all the states converge to the *average* of the initial states, this control is called the *average consensus control.* Also, since the local control is *distributed* over the network, we call the control scheme the *distributed control.*

Example 6.3. Let us consider the network in Fig. 6.1. Over this graph, we simulate the average consensus control in (6.11). Since the maximum degree is $\Delta = 3$, we choose the step size $\epsilon = 1/4$. We set the initial states as

$$x_1[0] = 10, \quad x_2[0] = 5, \quad x_3[0] = 0, \quad x_4[0] = -5, \quad x_5[0] = -10. \tag{6.17}$$

We note that the average of the initial states is given by $\overline{\boldsymbol{x}[0]} = 0$. Fig. 6.2 shows the state trajectories of $x_i[k]$, $i \in \{1, 2, 3, 4, 5\}$. We can see that all the states converge to the average value. The Python program for this simulation is given in Section 6.5.1 □

6.3 Distributed Optimization

In this section, we consider the following optimization problem:

$$\underset{\boldsymbol{x} \in \mathbb{R}^n}{\text{minimize}} \sum_{i=1}^{N} f_i(\boldsymbol{x}), \tag{6.18}$$

where $f_1, f_2, \ldots, f_N : \mathbb{R}^n \to \mathbb{R} \cup \{\infty\}$ are proper, closed, and convex functions. For this problem, we adapt the idea of consensus control to construct a *distributed optimization algorithm* with N agents where the i-th agent, $i \in \{1, 2, \ldots, N\}$, solves the following *local* optimization problem:

$$\underset{\boldsymbol{x} \in \mathbb{R}^n}{\text{minimize}} \ f_i(x). \tag{6.19}$$

The agents form a network $G = (\mathcal{V}, \mathcal{E})$ with $\mathcal{V} = \{1, 2, \ldots, N\}$ over which each agent exchanges its local computation results with neighboring agents.

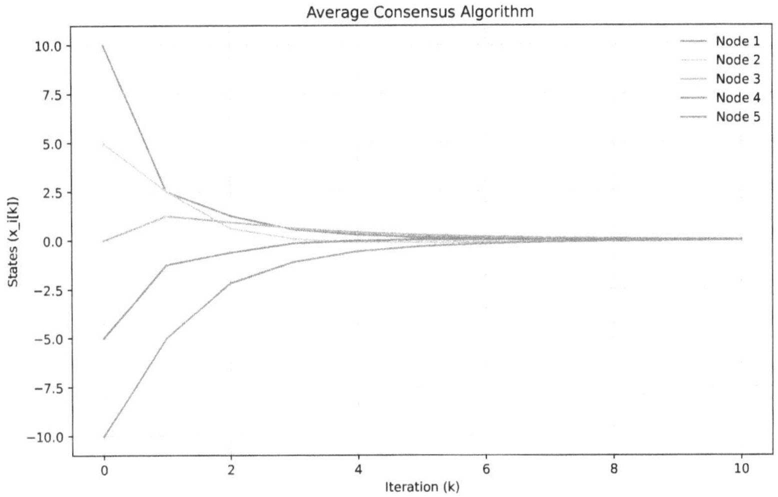

Figure 6.2: State trajectories of $x_i[k]$ with average consensus control.

6.3.1 Distributed gradient descent

In this section, we solve the optimization problem (6.18) by a distributed algorithm based on gradient descent over a network $G = (\mathcal{V}, \mathcal{E})$ with $\mathcal{V} = \{1, 2, \ldots, N\}$. Namely, each agent $i \in \mathcal{V}$ solves the local problem (6.19) and communicates local results with neighboring agents. We here assume G is undirected, simple, and connected, and $f_1, f_2, \ldots, f_N$ are differentiable.

First, the problem (6.18) can be solved by the *gradient descent algorithm* given by

$$\boldsymbol{x}[k+1] = \boldsymbol{x}[k] - \alpha \nabla f(\boldsymbol{x}[k]), \quad k = 0, 1, 2, \ldots, \tag{6.20}$$

where $f \triangleq f_1 + f_2 + \ldots + f_N$, and ∇f is the *gradient* of f defined by

$$\nabla f(\boldsymbol{x}) = \begin{bmatrix} \frac{\partial f}{\partial x_1}(\boldsymbol{x}) & \frac{\partial f}{\partial x_2}(\boldsymbol{x}) & \cdots & \frac{\partial f}{\partial x_n}(\boldsymbol{x}) \end{bmatrix}^\top. \tag{6.21}$$

Since the gradient is linear, we have

$$\boldsymbol{x}[k+1] = \boldsymbol{x}[k] - \alpha \sum_{i=1}^{N} \nabla f_i(\boldsymbol{x}[k]), \quad k = 0, 1, 2, \ldots \tag{6.22}$$

The idea of distributed gradient descent is to compute the local gradient $\nabla f_i(\boldsymbol{x}[k])$ by agent $i \in \mathcal{V}$. However, agent i can communicate only with its neighboring agents $j \in \mathcal{N}_i$, and the global variable $\boldsymbol{x}[k]$ is not directly accessible to each agent. Then, we adapt the consensus algorithm discussed in the previous section. Namely, we consider the following iteration for agent $i \in \mathcal{V}$:

$$\boldsymbol{x}_i[k+1] = \boldsymbol{x}_i[k] - \alpha \nabla f_i(\boldsymbol{x}_i[k]) + \boldsymbol{u}_i[k], \tag{6.23}$$

$$\boldsymbol{u}_i[k] = \epsilon \sum_{j \in \mathcal{N}_i} (\boldsymbol{x}_j[k] - \boldsymbol{x}_i[k]), \quad k = 0, 1, 2, \ldots. \tag{6.24}$$

We note that if we take $\boldsymbol{u}_i[k] = \boldsymbol{0}$, then the iteration is the gradient descent algorithm to find the minimizer of $f_i(\boldsymbol{x})$. Through the vector $\boldsymbol{u}_i[k]$, agent i obtains the partial information on the global variable $\boldsymbol{x}[k]$. Inserting (6.24) into (6.23) gives

$$\boldsymbol{x}_i[k+1] = \sum_{j=1}^{N} p_{ij} \boldsymbol{x}_j[k] - \alpha \nabla f_i(\boldsymbol{x}_i[k]), \tag{6.25}$$

where p_{ij} is the (i, j)-entry of the Perron matrix P_ϵ defined in (6.14). This is called the *distributed gradient descent algorithm*. For the convergence of this algorithm, see the paper [154].

Example 6.4. Let us consider a toy problem of minimizing the following cost function:

$$f(x) = \sum_{i=1}^{5} (x - i)^2. \tag{6.26}$$

This is a convex function, and the minimizer is $x = 3$, which can be found by

$$\nabla f(x) = \sum_{i=1}^{5} 2(x - i) = 10(x - 3) = 0. \tag{6.27}$$

The local cost function is $f_i(x) = (x - i)^2$ and its gradient (or derivative) is $\nabla f_i(x) = 2(x - i)$. The distributed gradient descent algorithm is then described as

$$x_i[k+1] = \sum_{j=1}^{5} p_{ij} x_j[k] - 2\alpha(x_i[k] - i), \quad k = 0, 1, 2, \ldots, \quad i = 1, 2, 3, 4, 5. \tag{6.28}$$

We use the network in Figure 6.1, and implement the distributed gradient descent algorithm (6.25) with Python. We set the initial states $x_i[0] = 0$

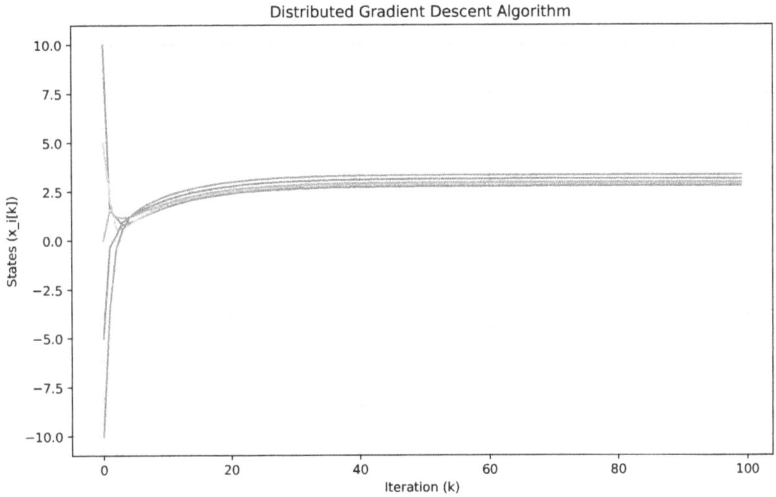

Figure 6.3: State trajectories of $x_i[k]$ with distributed gradient descent algorithm.

for all i, and the parameters $\epsilon = 1/4$ and $\alpha = 0.1$. Figure 6.3 shows the trajectories of $x_i[k]$. We can see that the states converge around the optimal value 3. Although there remain errors in the states, the average value $\overline{x[99]} = \frac{1}{5}\sum_{i=1}^{5} x_i[99] = 2.99992$ is close to the optimal value.

The Python program for this simulation is given in Section 6.5.2. □

6.3.2 Distributed ADMM algorithm

Here we form a distributed optimization problem for (6.18) based on ADMM (alternating direction method of multipliers; see Section 4.5, p. 90).

The optimization problem (6.18) is equivalently represented by

$$\underset{\boldsymbol{x}_1,\boldsymbol{x}_2,\ldots,\boldsymbol{x}_N\in\mathbb{R}^n}{\text{minimize}} \sum_{i=1}^{N} f_i(\boldsymbol{x}_i) \quad \text{subject to} \quad \boldsymbol{x}_1 = \boldsymbol{x}_2 = \cdots = \boldsymbol{x}_N. \tag{6.29}$$

Let us define the indicator function (see Section 4.2.4, p. 72) $I_{\mathcal{C}}$ with

$$\mathcal{C} \triangleq \{(\boldsymbol{x}_1, \boldsymbol{x}_2, \ldots, \boldsymbol{x}_N) : \boldsymbol{x}_1 = \boldsymbol{x}_2 = \cdots = \boldsymbol{x}_N\}. \tag{6.30}$$

The set $\mathcal{C}$ is a closed convex set, which is called the *consensus set*. Using this, we rewrite the optimization problem (6.29) as

$$\underset{\boldsymbol{x}_1,\boldsymbol{x}_2,\ldots,\boldsymbol{x}_N\in\mathbb{R}^n}{\text{minimize}} \sum_{i=1}^{N} f_i(\boldsymbol{x}_i) + I_{\mathcal{C}}(\boldsymbol{x}_1, \boldsymbol{x}_2, \ldots, \boldsymbol{x}_N). \tag{6.31}$$

Defining new variables $\boldsymbol{z}_i = \boldsymbol{x}_i$, $i = 1, 2, \ldots, N$, we have

$$\underset{\boldsymbol{x},\boldsymbol{z}\in\mathbb{R}^{nN}}{\text{minimize}} \ f(\boldsymbol{x}) + I_{\mathrm{C}}(\boldsymbol{z}) \ \text{ subject to } \ \boldsymbol{z} = \boldsymbol{x}, \tag{6.32}$$

where

$$\boldsymbol{x} \triangleq \left[\boldsymbol{x}_1^\top, \boldsymbol{x}_2^\top, \ldots, \boldsymbol{x}_N^\top\right]^\top \in \mathbb{R}^{nN}, \quad \boldsymbol{z} \triangleq \left[\boldsymbol{z}_1^\top, \ldots, \boldsymbol{z}_2^\top, \boldsymbol{z}_N^\top\right]^\top \in \mathbb{R}^{nN}, \tag{6.33}$$

and

$$f(\boldsymbol{x}) \triangleq \sum_{i=1}^{N} f_i(\boldsymbol{x}_i). \tag{6.34}$$

Now, the problem is equivalent to (4.99) with $f_1(\boldsymbol{x}) = f(\boldsymbol{x})$, $f_2(\boldsymbol{z}) = I_{\mathrm{C}}(\boldsymbol{z})$ and $\Psi = I$, and the associated ADMM algorithm is given by

$$\boldsymbol{x}[k+1] = \arg\min_{\boldsymbol{x}} \left\{ f(\boldsymbol{x}) + \frac{1}{2\gamma}\|\boldsymbol{x} - \boldsymbol{z}[k] + \boldsymbol{v}[k]\|_2^2 \right\}, \tag{6.35}$$

$$\boldsymbol{z}[k+1] = \Pi_{\mathcal{C}}\left(\boldsymbol{x}[k+1] + \boldsymbol{v}[k]\right), \tag{6.36}$$

$$\boldsymbol{v}[k+1] = \boldsymbol{v}[k] + \boldsymbol{x}[k+1] - \boldsymbol{z}[k+1]. \tag{6.37}$$

Then we will reformulate this iteration algorithm in a distributed algorithm. First, the objective function in (6.35) can be rewritten as

$$f(\boldsymbol{x}) + \frac{1}{2\gamma}\|\boldsymbol{x} - \boldsymbol{z}[k] + \boldsymbol{v}[k]\|_2^2 = \sum_{i=1}^{N} \left\{ f_i(\boldsymbol{x}_i) + \frac{1}{2\gamma}\|\boldsymbol{x}_i - \boldsymbol{z}_i[k] + \boldsymbol{v}_i[k]\|_2^2 \right\}. \tag{6.38}$$

Therefore, the update (6.35) for $\boldsymbol{x}[k+1]$ is divided into N updates as

$$\begin{aligned} \boldsymbol{x}_i[k+1] &= \arg\min_{\boldsymbol{x}_i} \{ f_i(\boldsymbol{x}_i) + \frac{1}{2\gamma}\|\boldsymbol{x}_i - \boldsymbol{z}_i[k] + \boldsymbol{v}_i[k]\|_2^2 \} \\ &= \operatorname{prox}_{\gamma f_i}(\boldsymbol{z}_i[k] - \boldsymbol{v}_i[k]), \quad i = 1, 2, \ldots, N, \end{aligned} \tag{6.39}$$

where we used the proximal operator (see Definition 4.4, p. 68).

For the projection $\Pi_{\mathcal{C}}$ onto the set $\mathcal{C}$ defined in (6.30) is given by the following lemma.

Lemma 6.1. The projection $\Pi_{\mathcal{C}}(\boldsymbol{x})$ with $\boldsymbol{x} = [\boldsymbol{x}_1^\top, \boldsymbol{x}_2^\top, \ldots, \boldsymbol{x}_N^\top]^\top$ is given by

$$\Pi_{\mathcal{C}}(\boldsymbol{x}) = \begin{bmatrix} \overline{\boldsymbol{x}} \\ \vdots \\ \overline{\boldsymbol{x}} \end{bmatrix}, \quad \overline{\boldsymbol{x}} \triangleq \frac{1}{N}\sum_{j=1}^{N} \boldsymbol{x}_j. \tag{6.40}$$

Using this lemma, we rewrite the update in (6.36) as

$$z_i[k+1] = \overline{x[k+1]+v[k]} = \frac{1}{N}\sum_{j=1}^{N}(x_j[k+1]+v_j[k]), \quad i=1,\dots,N. \tag{6.41}$$

Finally, from (6.37) and (6.41), we have

$$v_i[k+1] = v_i[k] + x_i[k+1] - \overline{x[k+1]} - \overline{v[k]}, \quad i=1,\dots,N. \tag{6.42}$$

On the other hand, we have

$$\begin{aligned}
\overline{v[k+1]} &= \frac{1}{N}\sum_{i=1}^{N} v_i[k+1]\\
&= \frac{1}{N}\sum_{i=1}^{N}\left\{v_i[k] + x_i[k+1] - \overline{x[k+1]} - \overline{v[k]}\right\}\\
&= \overline{v[k]} + \overline{x[k+1]} - \overline{x[k+1]} - \overline{v[k]}\\
&= \mathbf{0}.
\end{aligned} \tag{6.43}$$

If we set $v[0]$ such that $\overline{v[0]} = \mathbf{0}$, then we have $\overline{v[k]} = \mathbf{0}$ holds for $k = 0, 1, 2, \dots$, and we have the following distributed ADMM algorithm:

$$\begin{aligned}
x_i[k+1] &= \operatorname{prox}_{\gamma f_i}(\overline{x[k]} - v_i[k]),\\
v_i[k+1] &= v_i[k] + x_i[k+1] - \overline{x[k+1]}, \quad i=1,\dots,N.
\end{aligned} \tag{6.44}$$

The important point of this algorithm is that we can compute $x_i[k+1]$ locally by agent $i \in \{1, 2, \dots, N\}$ with global information $\overline{x[k]}$ given by a special agent called the *central collector* or the *fusion center*. The central collector collects data $x_i[k]$, $i = 1, \dots, N$, from all the agents, computes the average $\overline{x[k]}$, and sends it back to the agents. Figure 6.4 shows the network for the distributed ADMM algorithm. The central collector C connects all agents, and each agent ($i = 1, 2, 3$) only exchanges data with C. Such a network is called a *star network* as depicted in Figure 6.4.

6.3.3 Distributed least squares

As an application of the ADMM-based distributed optimization algorithm, we consider the problem of *least squares*. The optimization problem is minimizing the following objective function:

$$f(x) \triangleq \|Ax - b\|_2^2, \tag{6.45}$$

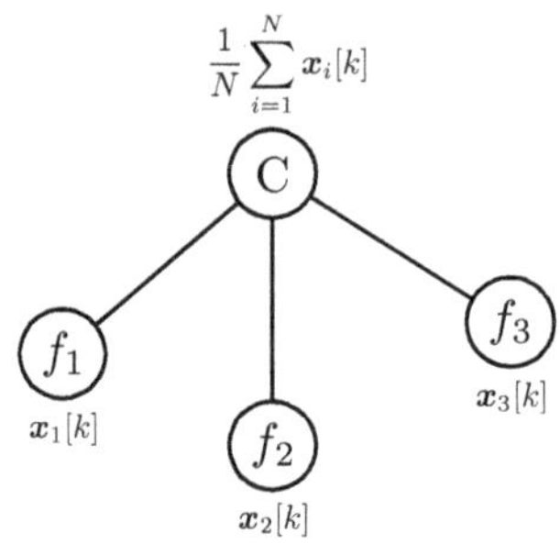

Figure 6.4: The network for distributed optimization (6.44).

where $A \in \mathbb{R}^{m\times n}$ and $\boldsymbol{b} \in \mathbb{R}^m$ are given. We assume that A has full column rank, that is, $\mathrm{rank}(A) = n$.

Now we solve the problem by distributed optimization. For this, let $A \triangleq [\boldsymbol{a}_1, \boldsymbol{a}_2, \ldots, \boldsymbol{a}_m]^\top$ with $\boldsymbol{a}_i \in \mathbb{R}^n$ and $\boldsymbol{b} \triangleq [b_1, b_2, \ldots, b_m]^\top$. Then we have

$$A\boldsymbol{x} - \boldsymbol{b} = \begin{bmatrix} \boldsymbol{a}_1^\top \boldsymbol{x} - b_1 \\ \boldsymbol{a}_2^\top \boldsymbol{x} - b_2 \\ \vdots \\ \boldsymbol{a}_m^\top \boldsymbol{x} - b_m \end{bmatrix}, \tag{6.46}$$

and hence

$$f(\boldsymbol{x}) = \|A\boldsymbol{x} - \boldsymbol{b}\|_2^2 = \sum_{i=1}^{m} (\boldsymbol{a}_i^\top \boldsymbol{x} - b_i)^2 = \sum_{i=1}^{m} f_i(\boldsymbol{x}), \tag{6.47}$$

where $f_i(\boldsymbol{x}) \triangleq (\boldsymbol{a}_i^\top \boldsymbol{x} - b_i)^2$. The proximal operator of f_i is derived as (see Section 4.2.3, p. 71)

$$\mathrm{prox}_{\gamma f_i}(\boldsymbol{v}) = \left(\boldsymbol{a}_i \boldsymbol{a}_i^\top + \gamma^{-1} I\right)^{-1} \left(b_i \boldsymbol{a}_i + \gamma^{-1}\boldsymbol{v}\right). \tag{6.48}$$

Finally, the ADMM-based distributed algorithm is given by

$$\begin{aligned} \boldsymbol{x}_i[k+1] &= M_i \left(b_i \boldsymbol{a}_i + \gamma^{-1}(\overline{\boldsymbol{x}[k]} - \boldsymbol{v}_i[k])\right), \\ \boldsymbol{v}_i[k+1] &= \boldsymbol{v}_i[k] + \boldsymbol{x}_i[k+1] - \overline{\boldsymbol{x}[k+1]}, \end{aligned} \tag{6.49}$$

where

$$M_i \triangleq \left(\boldsymbol{a}_i \boldsymbol{a}_i^\top + \gamma^{-1} I\right)^{-1}, i = 1, 2, \ldots, m. \tag{6.50}$$

The matrices $M_1, M_2, \ldots, M_m$ are fixed and can be computed beforehand to reduce the computational burden in the iteration.

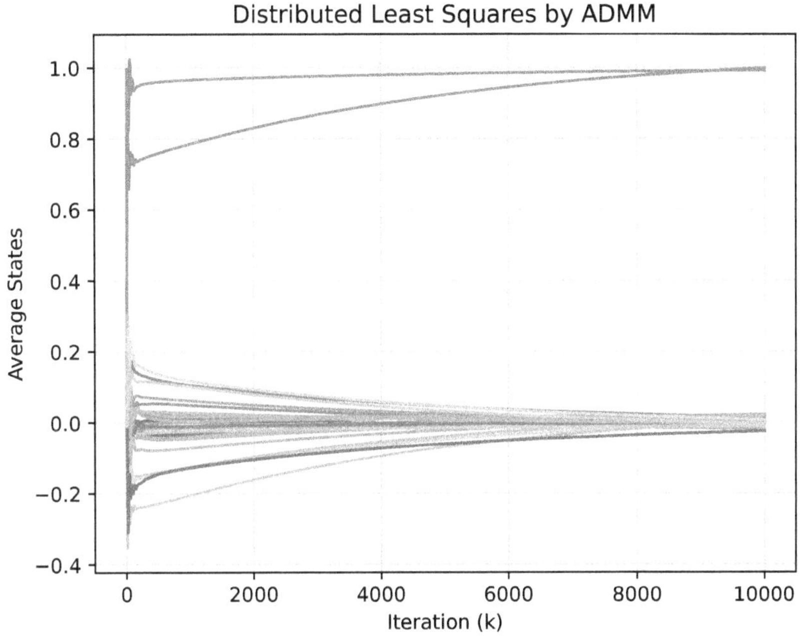

Figure 6.5: Average state $\overline{\boldsymbol{x}[k]}$, $k = 0, 1, 2, \ldots$ by distributed ADMM.

Example 6.5. We here execute an experiment of distributed least squares based on the ADMM-based algorithm (6.49). The matrix A is constructed as a binary-valued matrix of size 30×30. Let $\boldsymbol{x}_{\rm orig} \in \mathbb{R}^{30}$ be the original vector to be estimated, which is a sparse binary-valued vector with 2 non-zero elements. We note that $\|\boldsymbol{x}\|_0 = 2$. The vector $\boldsymbol{b} \in \mathbb{R}^{30}$ is computed by $\boldsymbol{b} = A\boldsymbol{x}_{\rm orig} + \boldsymbol{w}$ where $\boldsymbol{w} \in \mathbb{R}^{30}$ is a Gaussian noise vector with mean $\boldsymbol{0}$ and covariance $0.01^2 I$. We take the parameter γ in the ADMM-based algorithm (6.35) as $\gamma = 1$.

Figure 6.5 illustrates the convergence of the average state $\overline{\boldsymbol{x}[k]}$ over iterations $k = 0, 1, 2, \ldots$. As shown in this figure, the average $\overline{\boldsymbol{x}[k]}$ converges to the original vector $\boldsymbol{x}_{\rm orig}$, which consists of all zeros except for two elements with a value of one. Figure 6.6 shows the reconstructed vector $\overline{\boldsymbol{x}[10000]}$. While the reconstructed vector contains some errors, rounding it exactly recovers the original binary-valued vector.

The Python program for this simulation is shown in Section 6.5.3.

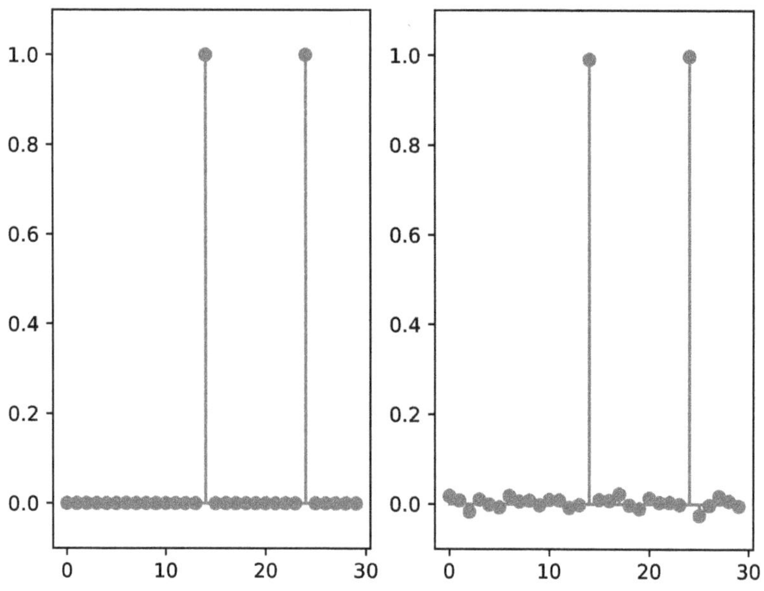

Figure 6.6: The original vector $\boldsymbol{x}_{\text{orig}}$ (left) and the reconstructed vector by distributed ADMM (right).

6.3.4 Distributed sparse regularization

Let us consider the following regularized optimization problem:

$$\underset{\boldsymbol{x}\in\mathbb{R}^n}{\text{minimize}} \ \sum_{i=1}^{N} f_i(\boldsymbol{x}) + g(\boldsymbol{x}), \tag{6.51}$$

where $g : \mathbb{R}^n \to \mathbb{R} \cup \{\infty\}$ is a proper, closed, and convex function such as $\|\boldsymbol{x}\|_1$ for sparse regularization. In this formulation, the regularization function g is assumed to be known to all agents.

The optimization problem (6.51) can be transformed into

$$\underset{\boldsymbol{x}_1,\boldsymbol{x}_2,\ldots,\boldsymbol{x}_N,\boldsymbol{z}\in\mathbb{R}^n}{\text{minimize}} \ \sum_{i=1}^{N} f_i(\boldsymbol{x}_i) + g(\boldsymbol{z}) \ \text{ subject to } \ \boldsymbol{x}_1 = \boldsymbol{x}_2 = \cdots = \boldsymbol{x}_N = \boldsymbol{z}. \tag{6.52}$$

Using the same idea used in Section 6.3.2, we have the following ADMM algorithm:

$$\boldsymbol{x}_i[k+1] = \text{prox}_{\gamma f_i}(\boldsymbol{z}[k] - \boldsymbol{v}_i[k]), \tag{6.53}$$
$$\boldsymbol{z}[k+1] = \text{prox}_{\gamma g/N}(\overline{\boldsymbol{x}[k+1]} + \overline{\boldsymbol{v}[k]}), \tag{6.54}$$
$$\boldsymbol{v}_i[k+1] = \boldsymbol{v}_i[k] + \boldsymbol{x}_i[k+1] - \boldsymbol{z}[k+1]. \tag{6.55}$$

In this distributed algorithm, the i-th agent locally updates the vectors $\boldsymbol{x}_i[k]$ and $\boldsymbol{v}_i[k]$ using the vector $\boldsymbol{z}[k]$, which is updated and transmitted by the central collector C as shown in Figure 6.4.

Example 6.6. Let us consider the following problem of *regularized least squares*:

$$\underset{\boldsymbol{x}}{\text{minimize}} \, \|A\boldsymbol{x} - \boldsymbol{b}\|_2^2 + \lambda\|\boldsymbol{x}\|_1. \tag{6.56}$$

This problem is for sparse vector reconstruction, and can be effectively solved by proximal gradient algorithms such as ISTA and FISTA as discussed in Section 4.4 (p. 81). We here solve this by a distributed optimization algorithm.

The distributed algorithm in (6.53)–(6.55) for (6.56) is given by

$$\boldsymbol{x}_i[k+1] = M_i\left(b_i\boldsymbol{a}_i + \gamma^{-1}(\boldsymbol{z}[k] - \boldsymbol{v}_i[k])\right), \tag{6.57}$$
$$\boldsymbol{z}[k+1] = S_{\gamma\lambda/N}\left(\overline{\boldsymbol{x}[k+1]} + \overline{\boldsymbol{v}[k]}\right), \tag{6.58}$$
$$\boldsymbol{v}_i[k+1] = \boldsymbol{v}_i[k] + \boldsymbol{x}_i[k+1] - \boldsymbol{z}[k+1], \tag{6.59}$$

where M_i is given in (6.50) and $S_{\gamma\lambda/N}$ is the soft-thresholding operator, the proximal operator of the ℓ^1 norm (see Section 4.2.5, p. 73).

Let A and $\boldsymbol{x}_{\text{orig}}$ be the same as in Example 6.5, and $\boldsymbol{b} = A\boldsymbol{x}_{\text{orig}} + \boldsymbol{w}$ where $\boldsymbol{w}$ is a Gaussian vector with mean $\boldsymbol{0}$ and covariance $0.01^2 I$. We take the regularization parameter $\lambda = 1$ in (6.56) and the step size $\gamma = 1$ for the ADMM algorithm (6.57)–(6.59). Figure 6.7 shows the average $\overline{\boldsymbol{x}[1000]}$ after 1000 iterations. The ℓ^1 norm in the regularization term promotes sparsity in the reconstructed vector, which contains smaller errors than that of the unregularized least squares discussed in Example 6.5.3.

The Python program in this simulation is given in Section 6.5.4. □

6.4 Further Readings

For the algebraic graph theory, the readers can refer to a nice book [5]. The consensus algorithm is the fundamental topic in the control of multi-agent systems, for which [17], [42], [98] are good references. In particular, the

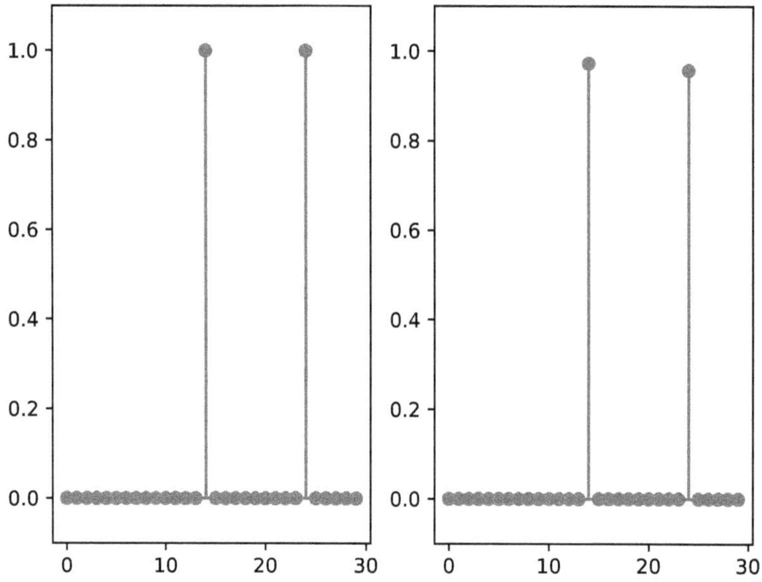

Figure 6.7: The original vector $\boldsymbol{x}_{\text{orig}}$ (left) and the reconstructed vector by distributed ADMM with regularization (right).

graph Laplacian plays an important role in multi-agent control systems [18].

For distributed optimization, you can refer to survey papers [20], [94], [116]. For the distributed gradient descent method, see papers [117], [154]. ADMM-based distributed optimization algorithms are explained in detail in [13].

6.5 Python Programs

6.5.1 Example 6.3

The following program is for the simulation of average consensus control shown in Example 6.3.

```
# Average consensus control

import networkx as nx
import numpy as np
```

```
import matplotlib.pyplot as plt

# Create a graph object
G = nx.Graph()

# Add vertices (nodes)
V = [1,2,3,4,5]
G.add_nodes_from(V)

# Add edges
E = [(1,2),(1,3),(1,4),(3,4),(2,5),(4,5)]
G.add_edges_from(E)

# Initialize the states of the nodes
x = np.array([10.0, 5.0, 0.0, -5.0, -10.0])

# Define the epsilon value
epsilon = 1 / 4

# Run the average consensus algorithm for a
   certain number of iterations
N_itr = 10
N_agents = len(V)
x_history = np.zeros([N_itr,N_agents]) # Store
   initial state
for k in range(N_itr):
  x_history[k,:] = x
  u = np.zeros_like(x)
  for i in range(len(V)):
    for j in range(len(V)):
      if (V[i], V[j]) in E or (V[j], V[i]) in E:
        u[i] += x[j] - x[i]
  u *= epsilon
  x += u

# Plot the time series for x_i[k]
plt.figure(figsize=(10, 6))
plt.plot(x_history)
```

```
plt.xlabel('Iteration (k)')
plt.ylabel('States (x_i[k])')
plt.title('Average Consensus Algorithm')
plt.grid(True)
plt.show()
```

6.5.2 Example 6.4

The following program is for the simulation of the distributed gradient descent algorithm shown in Example 6.4.

```
# Distributed Gradient Descent

import networkx as nx
import numpy as np
import matplotlib.pyplot as plt

# Create a graph object
G = nx.Graph()

# Add vertices (nodes)
V = [1,2,3,4,5]
G.add_nodes_from(V)

# Add edges
E = [(1,2),(1,3),(1,4),(3,4),(2,5),(4,5)]
G.add_edges_from(E)

# Initialize the states of the nodes
x = np.array([10.0, 5.0, 0.0, -5.0, -10.0])

# Define the epsilon value
epsilon = 1 / 4

# Step size of DGD
alpha = 0.1

# Run the average consensus algorithm for a
```

```
    certain number of iterations
N_itr = 100
N_agents = len(V)
x_history = np.zeros([N_itr,N_agents]) # Store
    initial state
v = np.array([1,2,3,4,5])
for k in range(N_itr):
  x_history[k,:] = x
  u = np.zeros_like(x)
  for i in range(len(V)):
    for j in range(len(V)):
      if (V[i], V[j]) in E or (V[j], V[i]) in E:
        u[i] += x[j] - x[i]
  u *= epsilon
  x += u - alpha * (x - v)

# Estimation
x_est = np.sum(x) / N_agents
error = x_est - 3
print(x_est)

# Plot the time series for x_i[k]
plt.figure(figsize=(10, 6))
plt.plot(x_history)
plt.xlabel('Iteration (k)')
plt.ylabel('States (x_i[k])')
plt.title('Distributed Gradient Descent Algorithm
    ')
plt.grid(True)
plt.show()
```

6.5.3 Example 6.5

The following program is for the simulation of distributed least squares by the ADMM-based algorithm shown in Example 6.5.

```
# Distributed least squares by the ADMM-based
    algorithm
```

```

import numpy as np
import numpy.linalg as LA
import matplotlib.pyplot as plt

# Parameter settings
# random seed
np.random.seed(1)
# matrix A
n = 30
m = n
A = np.random.randn(m, n)
# original vector (n-dimensional, k-sparse)
k = 2
x_orig = np.zeros(n)
S = np.random.randint(n,size=k)
x_orig[S] = 1

# vector b
b = A @ x_orig + 0.01 * np.random.randn(m)

# Distributed ADMM
# Parameters
gamma = 1
N_agents = m
M = np.zeros([n,n,N_agents])
for i in range(N_agents):
  fi = A[i,:].reshape(1,-1)
  M[:,:,i] = LA.inv(fi.T @ fi + np.eye(n) / gamma)
# Iterations
max_itr = 10000 # number of iterations
x = np.zeros([n, N_agents])
u = np.zeros([n, N_agents])
x_bar = np.sum(x,axis=1) / N_agents
x_history = np.zeros([max_itr,N_agents])
for k in range(max_itr):
  x_history[k,:] = x_bar
  for i in range(N_agents):
```

```
    qi = b[i] * A[i,:]
    x_next = M[:,:,i] @ (qi + (x_bar - u[:,i]) / gamma)
    x[:,i] = x_next
  x_bar = np.sum(x,axis=1) / N_agents
  for i in range(N_agents):
    u_next = u[:,i] + x[:,i] - x_bar
    u[:,i] = u_next

# Estimation
x_est = np.sum(x,axis=1) / N_agents

# Reconstructed vector
fig = plt.figure()
ax1 = fig.add_subplot(1, 2, 1)
ax1.stem(x_orig)
ax1.set_ylim(-0.1,1.1)
ax2 = fig.add_subplot(1, 2, 2)
ax2.stem(x_est)
ax2.set_ylim(-0.1,1.1)
plt.show()

# Average state trajectory
fig = plt.figure()
plt.plot(x_history)
plt.xlabel('Iteration (k)')
plt.ylabel('Average States')
plt.title('Distributed Least Squares by ADMM')
plt.grid(True)
plt.show()
```

6.5.4 Example 6.6

The following program is for the simulation of distributed least squares with sparse regularization shown in Example 6.6.

```
# Distributed sparse regularization
import cvxpy as cp
```

```
import numpy as np
import numpy.linalg as LA
import matplotlib.pyplot as plt
from mpl_toolkits.mplot3d import axes3d
import networkx as nx

## Parameter settings
# random seed
np.random.seed(1)
# matrix A
n = 30
m = n
A = np.random.randn(m, n)
# original vector (n-dimensional, k-sparse)
k = 2
x_orig = np.zeros(n)
S = np.random.randint(n,size=k)
x_orig[S] = 1

# vector b
b = A @ x_orig + 0.01 * np.random.randn(m)

# Optimization parameter settings
lmbd = 1

# Optimization by distributed ADMM
# Soft-thresholding function
def St(lmbd, v):
    n = v.shape[0]
    Sv = np.zeros(n)
    i = np.abs(v) > lmbd
    Sv[i] = v[i] - np.sign(v[i]) * lmbd
    return Sv

# Parameters
gamma = 1
N_agents = m

```

```
M = np.zeros([n,n,N_agents])
for i in range(N_agents):
  fi = A[i,:].reshape(1,-1)
  M[:,:,i] = LA.inv(fi.T @ fi + rho * np.eye(n))

Ap = np.linalg.pinv(A) # Moore-Penrose pseudo
   inverse of A

# Iterations
max_itr = 1000 # number of iterations
x = np.zeros([n, N_agents])
u = np.zeros([n, N_agents])
z = np.zeros(n)
x_bar = np.sum(x,axis=1) / N_agents
x_history = np.zeros([max_itr,N_agents])
for k in range(max_itr):
  x_history[k,:] = x_bar
  for i in range(N_agents):
    qi = b[i] * A[i,:]
    x_next = M[:,:,i] @ (qi + (z - u[:,i]) / gamma
       )
    x[:,i] = x_next
  x_bar = np.sum(x,axis=1) / N_agents
  u_bar = np.sum(u,axis=1) / N_agents
  z = St(gamma * lmbd/ N_agents, x_bar + u_bar)
  for i in range(N_agents):
    u_next = u[:,i] + x[:,i] - z
    u[:,i] = u_next

# Estimation
x_est = np.sum(x,axis=1) / N_agents

# Reconstructed vector
fig = plt.figure()
ax1 = fig.add_subplot(1, 2, 1)
ax1.stem(x_orig)
ax1.set_ylim(-0.1,1.1)
ax2 = fig.add_subplot(1, 2, 2)
```

```
ax2.stem(x_est)
ax2.set_ylim(-0.1,1.1)
plt.show()

# Average state trajectory
fig = plt.figure()
plt.plot(x_history)
plt.xlabel('Iteration (k)')
plt.ylabel('Average States')
plt.title('Distributed Sparse Regularization by
    ADMM')
plt.grid(True)
plt.show()
```

Chapter 7

Applications of Compressed Sensing

In this section, we showcase applications of compressed sensing and sparse optimization to systems and control.

7.1 Sparse Representations for Splines

Here we consider curve fitting by using splines. As in Chapter 3, we consider the following two-dimensional dataset:

$$\mathcal{D} = \{(t_1, y_1), (t_2, y_2), \ldots, (t_m, y_m)\}, \tag{7.1}$$

where $0 \leq t_1 < t_2 < \cdots < t_m = T$ represent the sampling times, and $y_1, y_2, \ldots, y_m$ are the corresponding observations. We assume that these observations are generated from the following model:

$$y_i = y(t_i) + \epsilon_i, \quad i = 1, 2, \ldots, m, \tag{7.2}$$

where y is a function and ϵ_i is additive noise. The problem of curve fitting is to estimate the unknown function y from the dataset $\mathcal{D}$.

In Chapter 3, we have assumed that the function is a polynomial function with a fixed order, and shown that the problem becomes a convex optimization. Here we seek a function among more general functions called *splines*. Namely, we consider the following optimization problem:

$$\underset{y}{\text{minimize}} \sum_{i=1}^{m} |y(t_i) - y_i|^2 + \lambda \int_0^T |\ddot{y}(t)|^2 dt, \tag{7.3}$$

where we assume the second derivative $\ddot{y}$ is in $L^2(0, T)$. The first term is for the fidelity of curve fitting to the data, and the second term is for the smoothness of the curve. In general, if you increase the fidelity then the curve becomes less smooth, and hence we need to control the trade-off between them to appropriately choose the parameter $\lambda > 0$.

Note that since y is not a finite-dimensional vector but a function, the problem is an *infinite-dimensional problem.* However, to use techniques in Hilbert space theory, the problem can be reduced to a finite-dimensional optimization problem. Let us first show this in this section. For this, we introduce the formulation of *control theoretic splines* [42], [145].

7.1.1 Solution by projection theorem

First, let us define

$$x_1(t) \triangleq y(t), \quad x_2(t) \triangleq \dot{y}(t), \quad u(t) \triangleq \ddot{y}(t). \tag{7.4}$$

Then, the optimization problem can be described as

$$\begin{aligned} \underset{u \in L^2(0,T)}{\text{minimize}} \quad & \sum_{i=1}^{m} |y(t_i) - y_i|^2 + \lambda \int_0^T |u(t)|^2 dt \\ \text{subject to} \quad & \dot{\boldsymbol{x}}(t) = A\boldsymbol{x}(t) + \boldsymbol{b}u(t), \quad y(t) = \boldsymbol{c}^\top \boldsymbol{x}(t), \quad t \in [0, T], \\ & \boldsymbol{x}(0) = \boldsymbol{0}, \end{aligned} \tag{7.5}$$

where $\boldsymbol{x}(t) \triangleq [x_1(t), x_2(t)]^\top$ and

$$A \triangleq \begin{bmatrix} 0 & 1 \\ 0 & 0 \end{bmatrix}, \quad \boldsymbol{b} \triangleq \begin{bmatrix} 0 \\ 1 \end{bmatrix}, \quad \boldsymbol{c} \triangleq \begin{bmatrix} 1 \\ 0 \end{bmatrix}. \tag{7.6}$$

Note that this formulation is for more general optimization than (7.3) by choosing another set of $(A, \boldsymbol{b}, \boldsymbol{c})$.

Define

$$l(\tau, t) \triangleq \begin{cases} \boldsymbol{c}^\top e^{A(t-\tau)}\boldsymbol{b}, & \text{if } 0 \le t \le \tau, \\ 0, & \text{otherwise}, \end{cases} \tag{7.7}$$

and

$$\phi_i(t) \triangleq l(t, t_i), \quad i = 1, 2, \ldots, m. \tag{7.8}$$

Then, we have

$$y(t_i) = \langle \phi_i, u \rangle_{L^2} = \int_0^T \phi_i(t) u(t) dt, \quad i = 1, 2, \ldots, m. \tag{7.9}$$

From this, the problem (7.5) becomes

$$\underset{u \in L^2(0,T)}{\text{minimize}} \sum_{i=1}^{m} \left| \langle \phi_i, u \rangle_{L^2} - y_i \right|^2 + \lambda \int_0^T |u(t)|^2 dt. \tag{7.10}$$

Then, if we define $z_i \triangleq \langle \phi_i, u \rangle_{L^2}$, the optimization problem is described as

$$\begin{aligned} &\underset{u \in L^2(0,T)}{\text{minimize}} \quad \sum_{i=1}^{m} |z_i - y_i|^2 + \lambda \int_0^T |u(t)|^2 dt \\ &\text{subject to} \quad z_i = \langle \phi_i, u \rangle_{L^2}, \quad i = 1, 2, \ldots, m. \end{aligned} \tag{7.11}$$

Define a new Hilbert space H by

$$H = L^2(0,T) \times \mathbb{R}^m, \tag{7.12}$$

with inner product

$$\left\langle \begin{bmatrix} v \\ \boldsymbol{w} \end{bmatrix}, \begin{bmatrix} u \\ \boldsymbol{z} \end{bmatrix} \right\rangle_H \triangleq \boldsymbol{w}^\top \boldsymbol{z} + \int_0^T v(t) u(t) dt. \tag{7.13}$$

Then, consider a closed linear subspace M of H defined by

$$M \triangleq \left\{ \begin{bmatrix} u \\ \boldsymbol{z} \end{bmatrix} \in H : z_i = \langle \phi_i, u \rangle_{L^2} \right\}, \tag{7.14}$$

and a vector $\boldsymbol{p} \in H$ defined by

$$\boldsymbol{p} \triangleq \begin{bmatrix} 0 \\ \boldsymbol{y} \end{bmatrix} \in H, \tag{7.15}$$

where $\boldsymbol{y} = [y_1, y_2, \ldots, y_m]^\top \in \mathbb{R}^m$. Then, for $\boldsymbol{r} \triangleq (u, \boldsymbol{z}) \in H$, we have

$$\|\boldsymbol{r} - \boldsymbol{p}\|_H^2 = \sum_{i=1}^{m} |z_i - y_i|^2 + \lambda \int_0^T |u(t)|^2 dt, \tag{7.16}$$

where $\|\cdot\|_H$ is the norm induced by the inner product $\langle \cdot, \cdot \rangle_H$, that is

$$\|\boldsymbol{r} - \boldsymbol{p}\|_H = \sqrt{\langle \boldsymbol{r} - \boldsymbol{p}, \boldsymbol{r} - \boldsymbol{p} \rangle_H}. \tag{7.17}$$

The optimization problem (7.11) is now rewritten as

$$\underset{r \in H}{\text{minimize}} \, \|\boldsymbol{r} - \boldsymbol{p}\|_H^2 \quad \text{subject to} \quad \boldsymbol{r} \in M. \tag{7.18}$$

The minimizer is given by the *projection* of $\boldsymbol{p} \in H$ onto the closed linear subspace $M \subset H$. Let $M^\perp$ denote the *orthogonal complement* of M in H. That is,

$$M^\perp \triangleq \left\{ \begin{bmatrix} v \\ \boldsymbol{w} \end{bmatrix} : \left\langle \begin{bmatrix} v \\ \boldsymbol{w} \end{bmatrix}, \begin{bmatrix} u \\ \boldsymbol{z} \end{bmatrix} \right\rangle_H = 0, \ \forall \begin{bmatrix} u \\ \boldsymbol{z} \end{bmatrix} \in M \right\}. \tag{7.19}$$

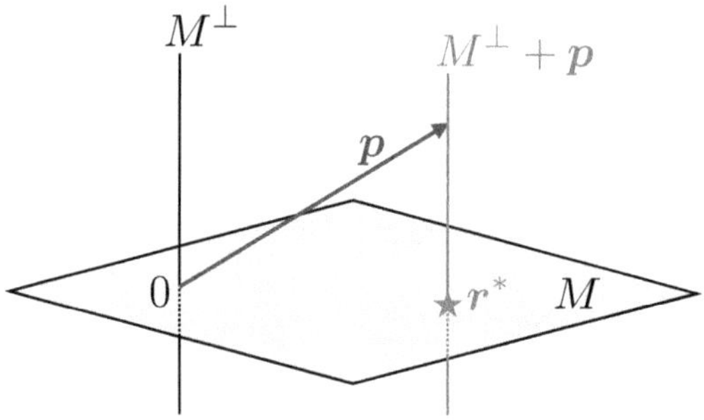

Figure 7.1: Projection theorem: the projection of $\boldsymbol{p}$ onto M is given by $\boldsymbol{r}^* \in (M^\perp + \boldsymbol{p}) \cap M$.

Then, from the projection theorem, the minimizer $\boldsymbol{r}^*$ is in the set $(M^\perp + \boldsymbol{p}) \cap M$ (see [42, Section 2.3]). Figure 7.1 illustrates this property from the projection theorem.

Now, let us characterize the set $M^\perp + \boldsymbol{p}$. Take $(v, \boldsymbol{w}) \in M^\perp$. Then, from (7.14), for any $(u, \boldsymbol{z}) \in M$, we have

$$\begin{aligned}
0 &= \left\langle \begin{bmatrix} v \\ \boldsymbol{w} \end{bmatrix}, \begin{bmatrix} u \\ \boldsymbol{z} \end{bmatrix} \right\rangle_H \\
&= \boldsymbol{w}^\top \boldsymbol{z} + \lambda \int_0^T v(t)u(t)dt \\
&= \boldsymbol{w}^\top \begin{bmatrix} \langle \phi_1, u \rangle_{L^2} \\ \langle \phi_2, u \rangle_{L^2} \\ \vdots \\ \langle \phi_m, u \rangle_{L^2} \end{bmatrix} + \lambda \langle v, u \rangle_{L^2} \\
&= \sum_{i=1}^m w_i \langle \phi_i, u \rangle_{L^2} + \lambda \langle v, u \rangle_{L^2} \\
&= \left\langle \sum_{i=1}^m w_i \phi_i + \lambda v, u \right\rangle_{L^2}.
\end{aligned} \tag{7.20}$$

This equation holds for any $u \in L^2(0, T)$, and hence we have

$$\sum_{i=1}^m w_i \phi_i + \lambda v = 0, \tag{7.21}$$

or

$$v = -\frac{1}{\lambda} \sum_{i=1}^m w_i \phi_i. \tag{7.22}$$

From this, the subspace $M^\perp$ can be represented by

$$M^\perp = \left\{ \begin{bmatrix} -\frac{1}{\lambda}\sum_{i=1}^m w_i\phi_i \\ \boldsymbol{w} \end{bmatrix} : \boldsymbol{w} \in \mathbb{R}^m \right\}, \tag{7.23}$$

and also we have

$$M^\perp + \boldsymbol{p} = \left\{ \begin{bmatrix} -\frac{1}{\lambda}\sum_{i=1}^m w_i\phi_i \\ \boldsymbol{w}+\boldsymbol{y} \end{bmatrix} : \boldsymbol{w} \in \mathbb{R}^m \right\}. \tag{7.24}$$

Now, let us obtain the minimizer $\boldsymbol{r}^* = (u^*, \boldsymbol{z}^*) \in (M^\perp + \boldsymbol{p}) \cap M$ of (7.18) (see also Figure 7.1). First, since $\boldsymbol{r}^* \in M$, we have

$$z_i^* = \langle \phi_i, u^* \rangle_{L^2}, \quad i = 1, 2, \dots, m. \tag{7.25}$$

Next, since $\boldsymbol{r}^* \in M^\perp + \boldsymbol{p}$, we have

$$u^* = -\frac{1}{\lambda}\sum_{i=1}^m w_i\phi_i, \tag{7.26}$$

$$z_i^* = w_i + y_i. \tag{7.27}$$

Inserting (7.26) into (7.25) gives

$$z_i^* = \left\langle \phi_i, -\frac{1}{\lambda}\sum_{j=1}^m w_j\phi_j \right\rangle_{L^2} = -\frac{1}{\lambda}\sum_{j=1}^m w_j \langle \phi_i, \phi_j \rangle_{L^2}. \tag{7.28}$$

From (7.27), we have

$$-\frac{1}{\lambda}\sum_{j=1}^m w_j \langle \phi_i, \phi_j \rangle_{L^2} = w_i + y_i, \tag{7.29}$$

or

$$(\lambda I + G)\boldsymbol{w} = -\lambda \boldsymbol{y}, \tag{7.30}$$

where G is the *Gram matrix* defined by

$$G \triangleq \begin{bmatrix} \langle \phi_1, \phi_1 \rangle & \langle \phi_1, \phi_2 \rangle & \dots & \langle \phi_1, \phi_m \rangle \\ \langle \phi_2, \phi_1 \rangle & \langle \phi_2, \phi_2 \rangle & \dots & \langle \phi_2, \phi_m \rangle \\ \vdots & \vdots & \ddots & \vdots \\ \langle \phi_m, \phi_1 \rangle & \langle \phi_m, \phi_2 \rangle & \dots & \langle \phi_m, \phi_m \rangle \end{bmatrix}. \tag{7.31}$$

Since $\lambda > 0$, the matrix $\lambda I + G$ is non-singular, and hence

$$\boldsymbol{w} = -\lambda(\lambda I + G)^{-1}\boldsymbol{y}. \tag{7.32}$$

Finally, from (7.26), we obtain the solution

$$\begin{aligned} u^* &= -\frac{1}{\lambda}\sum_{i=1}^{m}[-(\lambda I + G)^{-1}\lambda \boldsymbol{y}]_i \phi_i \\ &= \sum_{i=1}^{m}[(\lambda I + G)^{-1}\boldsymbol{y}]_i \phi_i \\ &= \sum_{i=1}^{m} \alpha_i^* \phi_i, \end{aligned} \tag{7.33}$$

where

$$\boldsymbol{\alpha}^* \triangleq \begin{bmatrix} \alpha_1^* \\ \vdots \\ \alpha_m^* \end{bmatrix} = (\lambda I + G)^{-1}\boldsymbol{y}. \tag{7.34}$$

The important point of the solution is that the optimal solution of the infinite-dimensional optimization problem in (7.5) is described as a finite number of *spline functions* $\phi_1, \ldots, \phi_m$ and the problem is reduced to computing the unknown coefficients $\alpha_1^*, \ldots, \alpha_m^*$. In other words, the original problem (7.5) is fundamentally a finite-dimensional optimization problem. Note that this property is generalized to the *representer theorem* in statistical machine learning [139].

Finally, the optimal solution y^* of the original optimization problem (7.3) is given by

$$y^*(t) = \int_0^t \int_0^\tau u^*(s) ds d\tau. \tag{7.35}$$

7.1.2 Sparse representation

From (7.33), the number of coefficients is equal to m, the number of data. If the data is very big (i.e., m is very large), then we need many coefficients to represent the fitting curve $y(t)$. Then, to use the idea of sparse representation, we can reduce the number of coefficients. For this, we restrict the feasible solutions of the optimization problem (7.10) to be

$$u(t) = \sum_{i=1}^{m} z_i \phi_i(t), \tag{7.36}$$

where $z_1, \ldots, z_m$ are unknown coefficients to be obtained. With this, we have

$$\langle \phi_i, u \rangle_{L^2} = \left\langle \phi_i, \sum_{j=1}^{m} z_j \phi_j \right\rangle_{L^2} = \sum_{j=1}^{m} z_j \langle \phi_i, \phi_j \rangle_{L^2}, \tag{7.37}$$

and hence

$$\begin{bmatrix} \langle \phi_1, u \rangle_{L^2} \\ \langle \phi_2, u \rangle_{L^2} \\ \vdots \\ \langle \phi_m, u \rangle_{L^2} \end{bmatrix} = G\boldsymbol{z}, \tag{7.38}$$

and

$$\sum_{i=1}^{m} \big| \langle \phi_i, u \rangle_{L^2} - y_i \big| = \|G\boldsymbol{z} - \boldsymbol{y}\|^2. \tag{7.39}$$

Also, we have

$$\begin{aligned} \lambda \int_0^T |u(t)|^2 dt &= \lambda \int_0^T \left(\sum_{i=1}^{m} z_i \phi_i(t) \right) \left(\sum_{j=1}^{m} z_j \phi_j(t) \right) dt \\ &= \lambda \sum_{i=1}^{m} \sum_{j=1}^{m} z_i z_j \langle \phi_i, \phi_j \rangle_{L^2} \\ &= \lambda \boldsymbol{z}^\top G \boldsymbol{z}. \end{aligned} \tag{7.40}$$

Therefore, under the assumption of (7.36), the optimization problem (7.10) is rewritten as

$$\underset{\boldsymbol{z} \in \mathbb{R}^m}{\text{minimize}} \ \|G\boldsymbol{z} - \boldsymbol{y}\|^2 + \lambda \boldsymbol{z}^\top G \boldsymbol{z}. \tag{7.41}$$

Then, to promote the sparsity of $\boldsymbol{z}$, we add the ℓ^0 norm as a regularization term:

$$\underset{\boldsymbol{z} \in \mathbb{R}^m}{\text{minimize}} \ \|G\boldsymbol{z} - \boldsymbol{y}\|^2 + \lambda \boldsymbol{z}^\top G \boldsymbol{z} + \rho \|\boldsymbol{z}\|_0, \tag{7.42}$$

where $\rho > 0$ is the regularization parameter. As usual, we can adopt the ℓ^1 norm as convex relaxation of the ℓ^0 norm. The relaxed convex optimization problem is described as follows:

$$\underset{\boldsymbol{z} \in \mathbb{R}^m}{\text{minimize}} \ \|G\boldsymbol{z} - \boldsymbol{y}\|^2 + \lambda \boldsymbol{z}^\top G \boldsymbol{z} + \rho \|\boldsymbol{z}\|_1. \tag{7.43}$$

This can be easily solved by the proximal gradient algorithm studied in Section 4.4 (p. 81).

7.2 Sparse System Identification

System identification is the process of constructing mathematical models of dynamical systems from input-output data. In system identification, we use the input-output data to estimate the characteristics of a system, such as its impulse response, transfer function, and state-space equations. In

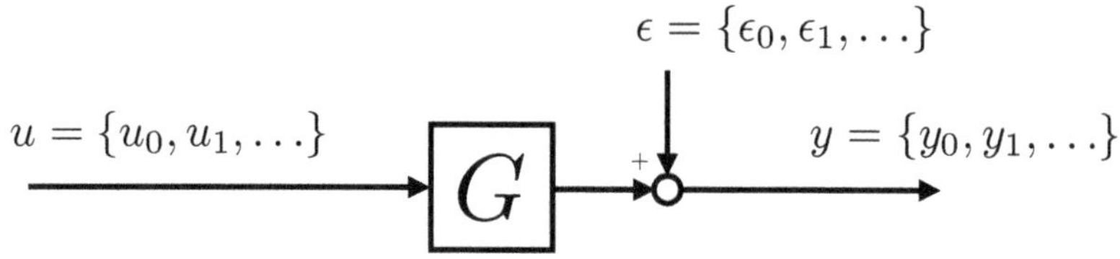

Figure 7.2: Block diagram

this section, we focus on a linear time-invariant discrete-time system. In particular, we focus on the *finite impulse response* (FIR) model, which is described as

$$y_k = \sum_{i=0}^{m-1} g_i u_{k-i} + \epsilon_k, \quad k = 0, 1, 2, \dots, N, \tag{7.44}$$

where $u = \{u_0, u_1, \dots, u_N\}$ and $y = \{y_0, y_1, \dots, y_N\}$ are respectively the input and output sequences, $\epsilon = \{\epsilon_0, \epsilon_1, \dots \epsilon_N\}$ is noise, and $g = \{g_0, g_1, \dots, g_{m-1}\}$ is the finite impulse response. In (7.44), we set $u_i = 0$ for $i < 0$. Also, we assume $N > m$. The block diagram of the system (7.44) is shown in Figure 7.2, where G denotes the linear system with impulse response g. The problem is to estimate the impulse response parameters $g_0, g_1, \dots, g_{m-1}$ from the input/output dataset

$$\mathcal{D} \triangleq \{(u_0, y_0), (u_1, y_1), \dots, (u_N, y_N)\}. \tag{7.45}$$

First, we define the following vectors:

$$\boldsymbol{y} \triangleq \begin{bmatrix} y_0 \\ y_1 \\ \vdots \\ y_N \end{bmatrix}, \quad \boldsymbol{\epsilon} \triangleq \begin{bmatrix} \epsilon_0 \\ \epsilon_1 \\ \vdots \\ \epsilon_N \end{bmatrix}, \quad \boldsymbol{g} \triangleq \begin{bmatrix} g_0 \\ g_1 \\ \vdots \\ g_{m-1} \end{bmatrix}. \tag{7.46}$$

Also, we define the following matrix called the *Toeplitz matrix*:

$$U \triangleq \begin{bmatrix} u_0 & 0 & \dots & 0 \\ u_1 & u_0 & \ddots & \vdots \\ u_2 & u_1 & \ddots & 0 \\ \vdots & \vdots & & u_0 \\ \vdots & \vdots & & \vdots \\ u_{N-1} & u_{N-2} & \dots & u_{N-m} \end{bmatrix} \in \mathbb{R}^{N \times m}. \tag{7.47}$$

We assume U has full column rank, that is, $\operatorname{rank}(U) = m$. Then, the relation (7.44) is rewritten as

$$\boldsymbol{y} = U\boldsymbol{g} + \boldsymbol{\epsilon}. \tag{7.48}$$

To estimate the vector $\boldsymbol{g}$, we consider the squared ℓ^2 error:

$$E(\boldsymbol{g}) \triangleq \frac{1}{2}\|\boldsymbol{y} - U\boldsymbol{g}\|_2^2. \tag{7.49}$$

The minimizer $\boldsymbol{g}_{\mathrm{LS}}$ that minimizes $E(\boldsymbol{g})$ is the *least squares solution*, which is given by (see Exercise 3.3, p. 42)

$$\boldsymbol{g}_{\mathrm{LS}} = (U^\top U)^{-1} U^\top \boldsymbol{y}. \tag{7.50}$$

Now, we further assume that the impulse response is sparse. For example, a delayed impulse response has zero coefficients for the first several steps and can thus be considered sparse. Applying the idea of ℓ^1 regularization discussed in Section 4.4, we minimize the following cost function:

$$J(\boldsymbol{g}) \triangleq E(\boldsymbol{g}) + \lambda\|\boldsymbol{g}\|_1 = \frac{1}{2}\|\boldsymbol{y} - U\boldsymbol{g}\|_2^2 + \lambda\|\boldsymbol{g}\|_1. \tag{7.51}$$

Minimizing this is easily done by the proximal gradient algorithm studied in Section 4.4.

Example 7.1. Here we consider a Python simulation of sparse system identification. We set the original impulse response to be identified is given by

$$g_i = \begin{cases} 1, & i = 5, 6, 7, 8, 9, \\ 0, & \text{otherwise.} \end{cases} \tag{7.52}$$

We set the length m of impulse response g as $m = 20$. The input u is a random sequence that takes values of ± 1. The length $N = 100$. Then, we generate the output y by (7.44) with Gaussian noise ϵ with mean 0 and variance $\sigma^2 = 0.1$. With these input/output data, we reconstruct the impulse response vector $\boldsymbol{g}$ by minimizing (7.51) with $\lambda = 5$. We adopt the FISTA algorithm (see 4.4.2, p. 83).

Figure 7.3 shows the original impulse response and the reconstructed impulse response. We can see that the sparse regularization accurately reconstructs the zero values in the impulse response. This property is useful when detecting the delay time in the response. Since the sparse regularization can accurately reconstruct the inactive portions of the response, it can precisely detect the delay time.

The Python program for this simulation is shown in Section 7.6.1. □

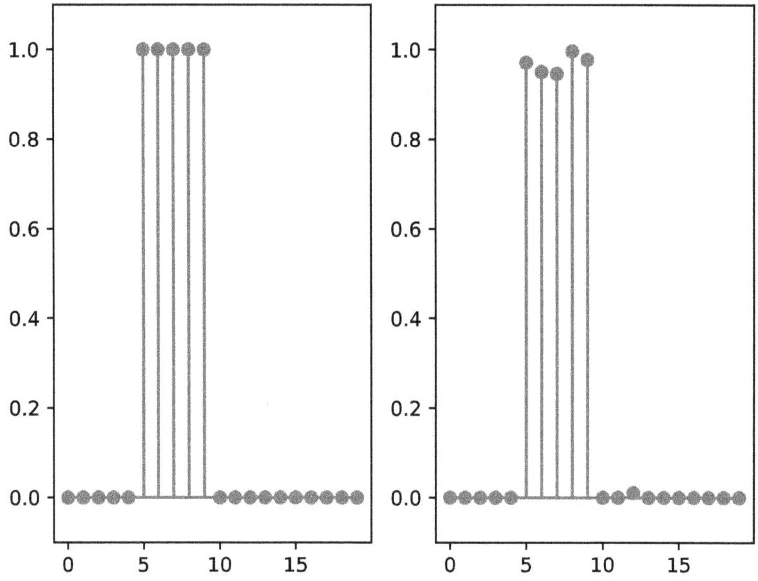

Figure 7.3: Original impulse response (left) and reconstructed response (right)

7.3 Sparse Controller Design

In this section, we consider stabilizing feedback control of continuous-time systems. We design a feedback controller that stabilizes an unstable system. In particular, we seek a *sparse* feedback gain matrix that has many zero entries.

Let us consider a linear time-invariant system described by the following *state equation*:

$$\dot{\boldsymbol{x}}(t) = A\boldsymbol{x}(t) + B\boldsymbol{u}(t), \quad t \geq 0, \tag{7.53}$$

where $\boldsymbol{x}(t) \in \mathbb{R}^d$, $\boldsymbol{u}(t) \in \mathbb{R}^m$, $A \in \mathbb{R}^{d\times d}$, and $B \in \mathbb{R}^{d\times m}$. We assume the pair (A, B) is *stabilizable*, or *asymptotically controllable* [142]. Then there exists a state feedback gain $K \in \mathbb{R}^{m\times d}$ such that the control

$$\boldsymbol{u}(t) = K\boldsymbol{x}(t), \tag{7.54}$$

asymptotically stabilizes the system (7.53), that is, the following property holds:

$$\lim_{t\to\infty} \boldsymbol{x}(t) = \mathbf{0}. \tag{7.55}$$

Inserting equation (7.54) into (7.53), we have

$$\dot{\boldsymbol{x}}(t) = (A + BK)\boldsymbol{x}(t), \tag{7.56}$$

and hence the matrix $A + BK$ is Hurwitz [142, Proposition 5.5.6].[1] This condition is equivalent to the existence of a positive definite matrix $Q \succ 0$ such that[2] the following inequality holds [141, Corollary 3.5.1]:

$$(A + BK)^\top Q + Q(A + BK) \prec 0. \tag{7.57}$$

The purpose is to obtain the feedback gain K, and we need to seek Q and K that satisfy the inequality. However, the products $K^\top Q$ and QK in (7.57) make the inequality nonlinear, which is difficult to solve directly. To address this, we introduce new variables $P \triangleq Q^{-1}$ and $Y \triangleq KP$. Then, from inequality (7.57), we derive the following inequalities:

$$P \succ 0, \quad AP + PA^\top + BY + Y^\top B^\top \prec 0. \tag{7.58}$$

These are called *linear matrix inequalities* (LMIs), which play an important role in the design of linear control systems [141].

If we find P and Y satisfying (7.58), then we can use Y as a stabilizing feedback gain with the *transformed* output $\boldsymbol{y}(t) = P^{-1}\boldsymbol{x}(t)$. Namely, the control input is given by $\boldsymbol{u}(t) = Y\boldsymbol{y}(t) = KP\boldsymbol{y}(t) = K\boldsymbol{x}(t)$. The problem of *sparse feedback gain* is to obtain a sparse Y that stabilizes the system. More precisely, we find matrix Y that has the minimum ℓ^0 norm, the number of nonzero elements in Y, among matrices satisfying the LMIs in (7.58).

To solve this, we slightly change the LMIs in (7.58) as follows:

$$P \succeq \epsilon I, \quad AP + PA^\top + BY + Y^\top B^\top \preceq -\epsilon I, \tag{7.59}$$

with a small number $\epsilon > 0$, and define

$$\Lambda \triangleq \{Y \in \mathbb{R}^{m \times d} : \exists P \succeq \epsilon I, AP + PA^\top + BY + Y^\top B^\top \preceq -\epsilon I\}. \tag{7.60}$$

We note that Λ is a closed subset of $\mathbb{R}^{m \times d}$, and if $Y \in \Lambda$ then this Y satisfies (7.58).

Now, our problem is formulated as the following optimization problem:

$$\underset{Y}{\text{minimize}} \, \|Y\|_0 \quad \text{subject to} \quad Y \in \Lambda. \tag{7.61}$$

[1] A matrix is said to be *Hurwitz* if all the eigenvalues of the matrix lie in the open left half plane in $\mathbb{C}$.

[2] We denote $Q \succ 0$ and $Q \succeq 0$ if Q is positive definite and positive semi-definite, respectively. Similarly, $Q \prec 0$ and $Q \preceq 0$ means Q is negative definite and negative semi-definite, respectively.

As always, we approximate the ℓ^0 norm of the matrix Y by the ℓ^1 norm $\|Y\|_1$, the sum of absolute values of the elements in Y. That is, we consider the following optimization problem:

$$\underset{Y}{\text{minimize}} \, \|Y\|_1 \text{ subject to } Y \in \Lambda. \tag{7.62}$$

We can adopt the Douglas-Rachford splitting algorithm (see Section 4.3.1, p. 77) to numerically solve this optimization problem. The algorithm is given by

$$\begin{aligned} Y[k+1] &= S_\gamma(Z[k]), \\ Z[k+1] &= Z[k] + \Pi_\Lambda(2Y[k+1] - Z[k]) - Y[k+1], \\ k &= 0, 1, 2, \ldots. \end{aligned} \tag{7.63}$$

The operator S_γ is the *soft-thresholding function* (see Section 4.2.5, p. 73) for a matrix defined by

$$[S_\gamma(V)]_{ij} \triangleq \begin{cases} V_{ij} - \gamma, & \text{if } V_{ij} \geq \gamma, \\ 0, & \text{if } -\gamma < V_{ij} < \gamma, \\ V_{ij} + \gamma, & \text{if } V_{ij} \leq -\gamma, \end{cases} \tag{7.64}$$

where $[S_\gamma(V)]_{ij}$ is the (i,j)-th entry of $S_\gamma(V) \in \mathbb{R}^{m \times d}$. The operator Π_Λ is the projection onto the set closed and convex Λ, which can be computed by solving another LMI optimization [12, Section 2.1] (see also [102]):

Lemma 7.1. For matrix $Y \in \mathbb{R}^{m \times d}$, the projection $\Pi_\Lambda(Y)$ is the solution of the following optimization problem:

$$\underset{S,Z}{\text{minimize}} \quad \text{trace}(S) \text{ subject to } \begin{bmatrix} S & (Z-Y)^\top \\ (Z-Y) & I \end{bmatrix} \succ 0, \; Z \in \Lambda. \tag{7.65}$$

We can also use a greedy method for the original problem (7.61). The *iterative greedy LMI* [102] is an alternating projection method to find an s-sparse matrix Y that satisfies $\|Y\|_0 \leq s$ in the LMI subset Λ for given $s \in \mathbb{N}$. The projection of matrix Y onto the subset of s-sparse matrices is given by

$$\arg\min_{Z} \|Z - Y\| \text{ subject to } \|Z\|_0 \leq s. \tag{7.66}$$

This projection is given by the *s-sparse operator* $\mathcal{H}_s(Y)$, which sets all but the s largest (in magnitude) elements of Y to 0 (see Section 5.3.2, p. 113).

The algorithm is given by

$$Y[k+1] = \mathcal{H}_s \circ \Pi_\Lambda(Y[k]), \quad k = 0, 1, 2, \dots, \tag{7.67}$$

with an initial guess $Y[0]$.

Example 7.2. Let us consider a linear system (7.53) with

$$A = \begin{bmatrix} 0 & 0 & 1.1320 & 0 & -1.0000 \\ 0 & -0.0538 & -0.1712 & 0 & 0.0705 \\ 0 & 0 & 0 & 1.0000 & 0 \\ 0 & 0.0485 & 0 & -0.8556 & -1.0130 \\ 0 & -0.2909 & 0 & 1.0532 & -0.6859 \end{bmatrix}, \tag{7.68}$$

$$B = \begin{bmatrix} 0 & 0 & 0 \\ -0.1200 & 1.0000 & 0 \\ 0 & 0 & 0 \\ 4.4190 & 0 & -1.6650 \\ 1.5750 & 0 & -0.0732 \end{bmatrix}. \tag{7.69}$$

This is the `AC1` model in the well-known benchmark problem set in COMPL$_e$ib library [81].

For this system, we design a sparse feedback gain by solving the ℓ^1 norm optimization in (7.62). We set $\epsilon = 0.01$ for (7.60). The optimal solution is given by

$$Y_{\text{alt}} = \begin{bmatrix} 0 & 0 & 0 & 0 & 0 \\ 0 & -0.00553 & 0 & 0 & 0 \\ 0 & 0 & 0 & 0 & 0 \end{bmatrix}. \tag{7.70}$$

This is sufficiently sparse, and we successfully obtain a sparse feedback gain.

The Python program to obtain the sparse feedback gain is shown in Section 7.6.2. □

7.4 Discrete-time Hands-off Control

In this section, we introduce sparse control (or hands-off control) for discrete-time systems.

7.4.1 Feasible control

Let us consider a linear time-invariant discrete-time system described by the following state equation:

$$\boldsymbol{x}[k+1] = A\boldsymbol{x}[k] + \boldsymbol{b}u[k], \quad k = 0, 1, 2, \dots, n-1, \tag{7.71}$$

where $\boldsymbol{x}[k] \in \mathbb{R}^d$ is the state and $u[k] \in \mathbb{R}$ is the control at time step $k \in \{0, 1, 2, \dots, n-1\}$. The matrix $A \in \mathbb{R}^{d \times d}$ and the vector $\boldsymbol{b} \in \mathbb{R}^d$ are assumed to be exactly known. The number n is the *horizon length* of the system.

Assume that the initial state $\boldsymbol{x}[0] = \boldsymbol{\xi}$ is given by state observation. Then the control objective is to find a control sequence $\{u[0], u[1], \dots, u[n-1]\}$ such that the control drives the state $\boldsymbol{x}[k]$ from $\boldsymbol{x}[0] = \boldsymbol{\xi}$ to the origin, that is,

$$\boldsymbol{x}[n] = \boldsymbol{0}. \tag{7.72}$$

From the state equation (7.71), we have

$$\boldsymbol{x}[k] = A^k \boldsymbol{x}[0] + \sum_{i=0}^{k-1} A^{k-1-i} \boldsymbol{b} u[i] = A^k \boldsymbol{\xi} + \sum_{i=0}^{k-1} A^{k-1-i} \boldsymbol{b} u[i], \tag{7.73}$$

for $k \in \{0, 1, \dots, n-1\}$. Then, the terminal constraint (7.72) can be described as

$$\boldsymbol{x}[n] = A^n \boldsymbol{\xi} + \sum_{i=0}^{n-1} A^{n-1-i} \boldsymbol{b} u[i] = A^n \boldsymbol{\xi} + \Phi \boldsymbol{u} = \boldsymbol{0}, \tag{7.74}$$

where

$$\Phi \triangleq \begin{bmatrix} A^{n-1}\boldsymbol{b} & A^{n-2}\boldsymbol{b} & \dots & A\boldsymbol{b} & \boldsymbol{b} \end{bmatrix}, \quad \boldsymbol{u} \triangleq \begin{bmatrix} u[0] \\ u[1] \\ \vdots \\ u[n-1] \end{bmatrix}. \tag{7.75}$$

We define the set of feasible controls that achieve (7.72) as

$$\mathcal{U}(n, \boldsymbol{\xi}) \triangleq \{\boldsymbol{u} \in \mathbb{R}^n : A^n \boldsymbol{\xi} + \Phi \boldsymbol{u} = \boldsymbol{0}\}. \tag{7.76}$$

For the feasibility, we have the following lemma.

Lemma 7.2. Suppose $n \geq d$ and the following matrix M is non-singular:

$$M \triangleq \begin{bmatrix} \boldsymbol{b} & A\boldsymbol{b} & \dots & A^{d-1}\boldsymbol{b} \end{bmatrix} \in \mathbb{R}^{d \times d}. \tag{7.77}$$

Then the feasible set $\mathcal{U}(n, \boldsymbol{\xi})$ is non-empty for any $\boldsymbol{\xi} \in \mathbb{R}^d$.

Note that the matrix M is called the *controllability matrix*, and the pair $(A, \boldsymbol{b})$ is called *controllable* if M is non-singular.

Proof of Lemma 7.2: Since $n \geq d$ and the matrix M is non-singular, the matrix Φ in (7.75) has full row rank. It follows that Φ is surjective and there exists at least one vector $\boldsymbol{u}$ that satisfies $\Phi \boldsymbol{u} = -A^n \boldsymbol{\xi}$ for any $\boldsymbol{\xi} \in \mathbb{R}^d$. □

7.4.2 Discrete-time maximum hands-off control

Now we consider *optimal control* that minimizes a cost function among control vectors in the feasible set $\mathcal{U}(n, \boldsymbol{\xi})$. A general form of the cost function is given by

$$J(\boldsymbol{u}) = \sum_{k=0}^{n-1} L(\boldsymbol{x}[k], u[k]), \tag{7.78}$$

where the function L is called the *stage cost function.*

The *linear quadratic control*, or *LQ control* for short, has the following stage cost function

$$L(\boldsymbol{x}, u) = \boldsymbol{x}^\top Q \boldsymbol{x} + r|u|^2, \tag{7.79}$$

where $Q \in \mathbb{R}^{d \times d}$ is a positive semidefinite matrix, and $r > 0$. In this section, we are interested in *sparse control*, also known as *maximum hands-off control*, which has the following stage cost function:

$$L(\boldsymbol{x}, u) = |u|^0. \tag{7.80}$$

With this, the cost function is given by

$$J(\boldsymbol{u}) = \sum_{k=0}^{n-1} |u[k]|^0 = \|\boldsymbol{u}\|_0. \tag{7.81}$$

The optimization problem is then described as

$$\underset{\boldsymbol{u} \in \mathbb{R}^n}{\text{minimize}} \ \|\boldsymbol{u}\|_0 \text{ subject to } \boldsymbol{u} \in \mathcal{U}(n, \boldsymbol{\xi}). \tag{7.82}$$

As usual, we approximate the ℓ^0 optimization by

$$\underset{\boldsymbol{u} \in \mathbb{R}^n}{\text{minimize}} \ \|\boldsymbol{u}\|_1 \text{ subject to } \boldsymbol{u} \in \mathcal{U}(n, \boldsymbol{\xi}), \tag{7.83}$$

which is the ℓ^1 optimization problem discussed in Section 4.3, and is efficiently solved by the Douglas-Rachford splitting algorithm (see Section 4.3.1).

Also one can consider the following cost function

$$J(\boldsymbol{u}) = \sum_{k=0}^{n-1} \left\{ \boldsymbol{x}[k]^\top Q \boldsymbol{x}[k] + \lambda |u[k]| \right\}, \tag{7.84}$$

with positive semidefinite $Q \in \mathbb{R}^{d\times d}$ and $\lambda > 0$. Inserting (7.73) into (7.84), we have

$$J(\boldsymbol{u}) = \boldsymbol{u}^\top R\boldsymbol{u} + 2\boldsymbol{q}^\top \boldsymbol{u} + \lambda\|\boldsymbol{u}\|_1 + c, \tag{7.85}$$

for some $R \in \mathbb{R}^{n\times n}$, $\boldsymbol{q} \in \mathbb{R}^n$, and $c \in \mathbb{R}$. For this optimization, we can apply the ADMM algorithm discussed in Section 4.5.

7.4.3 Model predictive control

As discussed above, the control sequence $\boldsymbol{u} \in \mathbb{R}^n$ is obtained by numerical optimization with a given initial state observation $\boldsymbol{\xi} \in \mathbb{R}^d$. Let $\mathcal{C}$ denote the mapping from the initial state $\boldsymbol{\xi} \in \mathbb{R}^d$ to the optimal control sequence $\boldsymbol{u} \in \mathbb{R}^n$, that is,

$$\boldsymbol{u} = \mathcal{C}(\boldsymbol{\xi}). \tag{7.86}$$

Then $\boldsymbol{u} = \mathcal{C}(\boldsymbol{\xi})$ is a finite-horizon control (i.e., the control is applied to a plant in a finite length of time), and this is *open-loop control.* Open-loop control is something like riding a bicycle with your eyes closed, which is very fragile against disturbances. To make the control system *robust,* you need to implement the control as *feedback control,* where the controller constantly observes the state and updates the control based on the latest state observation.

To implement feedback control from the finite-horizon control $\boldsymbol{u} = \mathcal{C}(\boldsymbol{\xi})$, we adopt the *model predictive control* (also known as *receding horizon control*). The model predictive control is described as follows:

1. Observe the state $\boldsymbol{x}[k]$ at time k.
2. Compute the optimal control sequence

$$\boldsymbol{u}[k] = \begin{bmatrix} u_0[k] \\ u_1[k] \\ \vdots \\ u_{n-1}[k] \end{bmatrix} = \mathcal{C}(\boldsymbol{x}[k]). \tag{7.87}$$

3. Use the first element of $\boldsymbol{u}[k]$, that is, $u_0[k]$, as the control at time k.

From this, the control $u[k]$ to the discrete-time plant (7.71) is obtained by

$$u[k] = u_0[k] = \begin{bmatrix} 1 & 0 & \dots & 0 \end{bmatrix} \mathcal{C}(\boldsymbol{x}[k]). \tag{7.88}$$

Figure 7.4 shows the block diagram of the feedback control system where $\mathcal{P}$ is the plant given in (7.71).

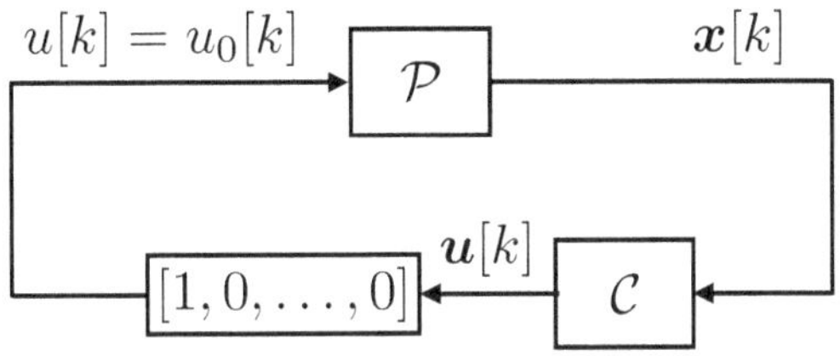

Figure 7.4: Feedback control by model predictive control $\boldsymbol{u}[k] = \mathcal{C}(\boldsymbol{x}[k])$. $\mathcal{P}$ is the plant given by (7.71).

The important thing we should do next is to study the *stability* of the feedback system. The closed-loop system in Figure 7.4 may exhibit instability, that is, $\boldsymbol{x}[k]$ may diverge, if we do not care about the stability. The instability is possible even when $\mathcal{P}$ and $\mathcal{C}$ are both stable. Therefore, to prove the stability is very important to design a feedback control system.

First, we define the *value function* $V(\boldsymbol{\xi})$ of the optimal control problem (7.83) by

$$V(\boldsymbol{\xi}) \triangleq \min_{u \in \mathcal{U}(n, \boldsymbol{\xi})} \|\boldsymbol{u}\|_1. \tag{7.89}$$

We have the following lemma:

Lemma 7.3. Assume that the controllability matrix M in (7.77) and the matrix A in (7.71) are non-singular. Assume also that $n \geq d$. Then the value function $V(\boldsymbol{\xi})$ is convex, continuous, and positive definite.

Exercise 7.1. Prove Lemma 7.3.

Now, we give a detailed definition of stability.

Definition 7.1. Let us consider the following discrete-time system

$$\boldsymbol{x}[k+1] = f(\boldsymbol{x}[k]), \quad k = 0, 1, 2, \ldots \tag{7.90}$$

Suppose that there exists the unique sequence $\{\boldsymbol{x}[0], \boldsymbol{x}[1], \ldots\}$ satisfying (7.90) for any initial state $\boldsymbol{x}[0] \in \mathbb{R}^d$. Suppose also that the origin is an equilibrium of the system, namely, $f(\boldsymbol{0}) = \boldsymbol{0}$ holds. Then the origin is said to be *stable* if for each $\epsilon > 0$ there exists $\delta > 0$ such that

$$\|\boldsymbol{x}[0]\|_2 < \delta \Rightarrow \|\boldsymbol{x}[k]\|_2 < \epsilon, \quad \forall k \geq 0. \tag{7.91}$$

The concept is very simple; the state trajectory $\{\boldsymbol{x}[k]\}_{k=0}^{\infty}$ starting out near the origin will keep on staying near the origin and never diverge.

From (7.71) and (7.88), the closed-loop system is described as

$$\boldsymbol{x}[k+1] = A\boldsymbol{x}[k] + \begin{bmatrix} 1 & 0 & \dots & 0 \end{bmatrix} \mathcal{C}(\boldsymbol{x}[k]) \triangleq f(\boldsymbol{x}[k]). \tag{7.92}$$

It is easily shown that the origin $\mathbf{0}$ is an equilibrium of this difference equation. To show the stability of this equilibrium, *Lyapunov's theorem* is available.

Theorem 7.1. Suppose that there exists a function $V : \mathbb{R}^d \to \mathbb{R}$ satisfying

1. $V(\mathbf{0}) = 0$.
2. $V(\boldsymbol{\xi})$ is continuous.
3. $V(\boldsymbol{\xi}) > 0$ for any $\boldsymbol{\xi} \neq \mathbf{0}$.
4. $V(\boldsymbol{x}[k+1]) \leq V(\boldsymbol{x}[k])$ for $k = 0, 1, 2, \dots$, for the state trajectory $\{\boldsymbol{x}[k]\}_{k=0}^{\infty}$ of the system (7.90).

Then the origin $\mathbf{0}$ is stable under the system equation (7.90).

A function V in Theorem 7.1 is called a *Lyapunov function.* The idea to prove the stability of our system (7.92) is to show the value function (7.89) to be a Lyapunov function. In fact, it is a Lyapunov function and we have the following theorem.

Theorem 7.2. Assume M and A are non-singular, and $n \geq d$. Then the origin is stable under the system equation (7.92).

Proof: We prove the value function $V(\boldsymbol{\xi})$ in (7.89) is a Lyapunov function of (7.92). The properties 1 to 3 in Theorem 7.1 are directly from Lemma 7.3. We here prove 4. Let

$$\boldsymbol{u}^*[k] \triangleq \begin{bmatrix} u_0^*[k] & u_1^*[k] & \dots & u_{n-1}^*[k] \end{bmatrix}^\top = \mathcal{C}(\boldsymbol{x}[k]), \tag{7.93}$$

and define

$$\tilde{\boldsymbol{u}}[k] \triangleq \begin{bmatrix} u_1^*[k] & u_2^*[k] & \dots & u_{n-1}^*[k], 0 \end{bmatrix}^\top. \tag{7.94}$$

Note that $\tilde{\boldsymbol{u}}[k]$ is a shifted control sequence by one time step of the optimal control sequence $\boldsymbol{u}^*[k]$ at time k. It is then easily shown that $\tilde{\boldsymbol{u}}[k]$ is a feasible control for $\boldsymbol{x}[k+1]$, that is,

$$\tilde{\boldsymbol{u}}[k] \in \mathcal{U}(n, \boldsymbol{x}[k+1]). \tag{7.95}$$

In fact, since $\boldsymbol{u}^*[k] \in \mathcal{U}(n, \boldsymbol{x}[k])$ we have

$$\begin{aligned}
&A^n \boldsymbol{x}[k+1] + \Phi\tilde{\boldsymbol{u}}[k] \\
&\quad = A^n(A\boldsymbol{x}[k] + \boldsymbol{b}u_0^*[k]) + A^{n-1}\boldsymbol{b}u_1^*[k] + \cdots + A\boldsymbol{b}u_{n-1}^*[k] \\
&\quad = A(A^n\boldsymbol{x}[k] + A^{n-1}\boldsymbol{b}u_0^*[k] + A^{n-2}\boldsymbol{b}u_1^*[k] + \cdots + \boldsymbol{b}u_{n-1}^*[k]) \\
&\quad = A(A^n\boldsymbol{x}[k] + \Phi\boldsymbol{u}^*[k]) \\
&\quad = A \times \boldsymbol{0} \\
&\quad = \boldsymbol{0}.
\end{aligned} \tag{7.96}$$

Then, from the optimality of the value function, we have

$$\begin{aligned}
V(\boldsymbol{x}[k+1]) &= \min\{\|\boldsymbol{u}\|_1 : \boldsymbol{u} \in \mathcal{U}(n, \boldsymbol{x}[k+1])\} \\
&\leq \|\tilde{\boldsymbol{u}}[k]\|_1 \\
&= |u_1^*[k]| + |u_2^*[k]| + \cdots + |u_{n-1}^*[k]| + |0| \\
&= \sum_{i=0}^{n-1} |u_i^*[k]| - |u_0^*[k]| \\
&= V(\boldsymbol{x}[k]) - |u_0^*[k]| \\
&\leq V(\boldsymbol{x}[k]),
\end{aligned} \tag{7.97}$$

for $k = 0, 1, 2, \ldots$ □

7.5 Further Readings

The control theoretic smoothing spline was first proposed in [145]. The book [42] is a nice reference for the smoothing spline. The convex optimization formulation of the constrained smoothing spline was considered in [99], and the sparse representation was proposed in [100].

The book [85] by L. Ljung covers fundamental concepts in system identification. Regularized system identification is discussed in [126]. A non-convex optimization approach is found in [111].

For the design of sparse feedback control, see papers [83], [84], [96], [102], [127]. The trade-off property between the closed-loop performance and the sparsity of the gain is discussed in [74]. A comprehensive survey of sparse feedback control can be found in recent books [72], [73].

The maximum hands-off control was first proposed in [105] for continuous-time and discrete-time systems. Detailed discussions of maximum hands-off control for continuous-time systems can be found in Part II of this book. The model predictive control formulation was proposed in [113].

7.6 Python Programs

7.6.1 Example 7.1

The following program is for the simulation of sparse system identification in Example 7.1.

```
import numpy as np
from scipy.linalg import toeplitz
from numpy.linalg import inv
import matplotlib.pyplot as plt

# input u
np.random.seed(0)
N = 100
u = np.random.rand(N)
u = np.where(u >= 0.5, 1, -1)

# true impulse response g*
m = 20
gstar = np.zeros([m,1])
gstar[5:10] = 1

# output y
sigma2 = 0.1 # noise sd
y = np.convolve(gstar.ravel(), u.ravel(), mode='
    full')[:N] + np.sqrt(sigma2)*np.random.randn(N)

# Toeplitz matrix U
U = toeplitz(np.concatenate(([u[0]], np.zeros(m -
    1))), u).T

## Optimization by FISTA
# Soft-thresholding function
def St(lmbd, v):
    n = v.shape[0]
```

```
    Sv = np.zeros(n)
    i = np.abs(v) > lmbd
    Sv[i] = v[i] - np.sign(v[i]) * lmbd
    return Sv

# parameter settings
lmbd = 5
U_norm = np.linalg.norm(U,2)
gamma = 1/U_norm**2 # step size
max_itr = 100 # number of iterations
g = np.zeros(m) # initial guess for g
z = g # initial guess for z
t = 0 # initial guess for t

# FISTA iteration
for k in range(max_itr):
  res = U @ z - y
  g2 = St(gamma*lmbd, z - gamma*U.T @ res)
  t2 = (1 + np.sqrt(1+4*t**2))/2
  z = g2 + (t-1)/t2 * (g2 - g)
  g = g2
  t = t2

# Plot the results
fig = plt.figure()
ax1 = fig.add_subplot(1, 2, 1)
ax1.stem(gstar)
ax1.set_ylim([-0.1,1.1])
ax2 = fig.add_subplot(1, 2, 2)
ax2.stem(g)
ax2.set_ylim([-0.1,1.1])
plt.show()
```

7.6.2 Example 7.2

The following program is for the design of a sparse feedback gain shown in Example 7.2.

```
# Import packages.
import cvxpy as cp
import numpy as np

# System matrices
n = 5
m = 3
A = np.matrix(
  [[0,0,1.1320,0,-1],
   [0,-0.0538,-0.1712,0,0.0705],
   [0,0,0,1,0],
   [0,0.0485,0,-0.8556,-1.0130],
   [0,-0.2909,0,1.0532,-0.6859]]
   )
B = np.matrix(
    [[0,0,0],
     [-0.12,1,0],
     [0,0,0],
     [4.419,0,-1.6650],
     [1.575,0,-0.0732]])

# LMIs
epsil = 0.01
eI = epsil * np.eye(n)
P = cp.Variable((n,n), symmetric=True)
Y = cp.Variable((m,n))
objective = cp.Minimize(cp.norm1(Y))
constraints = [P - eI >> 0]
constraints += [A @ P + P @ A.T + B @ Y + Y.T @ B.T + eI << 0]

# Optimization
prob = cp.Problem(objective, constraints)
prob.solve()
Y_ = Y.value
Y_[np.abs(Y_) < 1e-6] = 0

# Print result.
```

```
print("The optimal value is", prob.value)
print("A solution Y is")
print(Y_)
```

Part II

Maximum Hands-off Control: Compressed Sensing for Continuous-time Systems

Chapter 8

Dynamical Systems and Optimal Control

We have studied the idea, algorithms, and applications of compressed sensing in Part I. In Part II, we will extend the method to continuous-time dynamical systems. For this, we here review the basics of dynamical systems and optimal control.

Key ideas of Chapter 8

- A dynamical system is modeled by a differential equation called the state-space equation.
- We cannot control uncontrollable systems.
- Optimal control is the best control among feasible controls for a controllable system.

8.1 Dynamical Systems

A *dynamical system* is a system whose state changes over time. That is, a dynamical system is a *moving* system. Dynamical systems are all around us. They encompass a wide range of phenomena, from engineered systems like vehicles, airplanes, motors, and electric circuits, to natural phenomena like the movement of planets, changes in weather, ant swarms, and cell movement. Even social and economic systems, such as fluctuations in stock prices and the spread of viruses, can be modeled as dynamical systems.

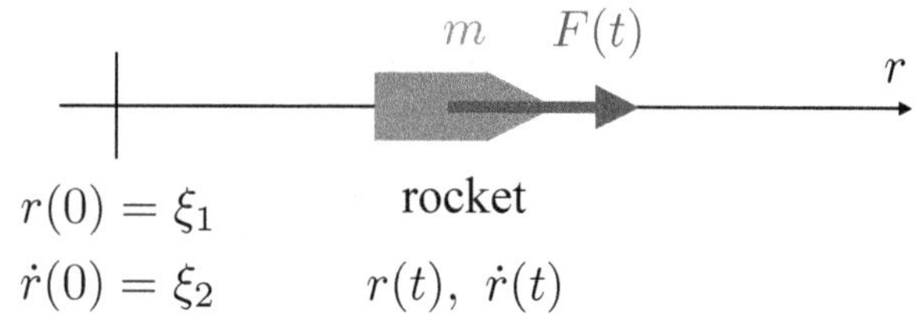

Figure 8.1: Rocket example

8.1.1 State equation

In Part II, we focus on a dynamical system that is described by a linear differential equation:

$$\dot{\boldsymbol{x}}(t) = A\boldsymbol{x}(t) + \boldsymbol{b}u(t), \quad t \geq 0, \quad \boldsymbol{x}(0) = \boldsymbol{\xi} \in \mathbb{R}^d, \tag{8.1}$$

where $A \in \mathbb{R}^{d\times d}$, $\boldsymbol{b} \in \mathbb{R}^d$, $\boldsymbol{x}(t) \in \mathbb{R}^d$, and $u(t) \in \mathbb{R}$. We call $\boldsymbol{x}(t)$ the *state*, and $u(t)$ the *control*. The state $\boldsymbol{x}(0) = \boldsymbol{\xi}$ at time $t = 0$ is called the *initial state*, and the differential equation in (8.1) is called the *state equation*. The dynamical system in (8.1) is controlled by the control $u(t)$, and is called a *controlled object* or a *plant*.

Exercise 8.1. Show that the solution of the differential equation (8.1) is given by

$$\boldsymbol{x}(t) = e^{At}\boldsymbol{\xi} + \int_0^t e^{A(t-\tau)}\boldsymbol{b}u(\tau)d\tau, \quad t \geq 0. \tag{8.2}$$

Example 8.1 (Rocket). Let us consider the control of a rocket in outer space where no friction nor gravity acts (see Figure 8.1). The rocket is accelerated by thrust from a rocket engine. Let the mass of the rocket be m. We assume that the rocket can move on a 1-dimensional straight line. Let the position of the rocket at time $t \geq 0$ be $r(t)$ with initial position $r(0) = \xi_1$, and initial velocity $v(0) = \dot{r}(0) = \xi_2$. We denote the thrust force by $F(t)$. From Newton's second law of motion, we have[1]

$$m\ddot{r}(t) = F(t), \quad r(0) = \xi_1, \quad \dot{r}(0) = \xi_2. \tag{8.3}$$

Let us transform this differential equation into the state equation in (8.1). For this, define the state $\boldsymbol{x}(t)$ by

$$\boldsymbol{x}(t) \triangleq \begin{bmatrix} x_1(t) \\ x_2(t) \end{bmatrix} \triangleq \begin{bmatrix} r(t) \\ \dot{r}(t) \end{bmatrix}. \tag{8.4}$$

[1]Strictly speaking, the thrust of a rocket is obtained by emitting its mass (e.g., fuel) to the opposite direction, and hence the model is not correct. That is, the mass m should be time-varying $m(t)$ that decreases in time. In this example, however, we assume that the mass of the rocket is sufficiently large and the variation can be ignored.

Then we have

$$\dot{\boldsymbol{x}}(t) = \begin{bmatrix} \dot{r}(t) \\ \ddot{r}(t) \end{bmatrix} = \begin{bmatrix} x_2(t) \\ m^{-1}F(t) \end{bmatrix} = \begin{bmatrix} 0 & 1 \\ 0 & 0 \end{bmatrix} \begin{bmatrix} x_1(t) \\ x_2(t) \end{bmatrix} + \begin{bmatrix} 0 \\ m^{-1} \end{bmatrix} F(t) \tag{8.5}$$

Defining $u(t) \triangleq F(t)$ and

$$A \triangleq \begin{bmatrix} 0 & 1 \\ 0 & 0 \end{bmatrix}, \quad \boldsymbol{b} \triangleq \begin{bmatrix} 0 \\ m^{-1} \end{bmatrix}, \quad \boldsymbol{\xi} \triangleq \begin{bmatrix} \xi_1 \\ \xi_2 \end{bmatrix}, \tag{8.6}$$

we obtain the state equation of the form (8.1). □

The system (8.3) or (8.5) is sometimes called the *double integrator*, since the position $r(t)$ is obtained by integrating $u(t) = F(t)$ twice.

Let us investigate the meaning of the state-space equation (8.1). Assume that the initial state $\boldsymbol{x}(0) = \boldsymbol{\xi}$ at time $t = 0$ is obtained from observation by a sensor attached to the system. Then we design $u(t)$ for $t \geq 0$ to realize a desired trajectory of the state $\boldsymbol{x}(t)$. In the rocket control considered in Example 8.1, we design the thrust force $u(t) = F(t)$ to drive the rocket, for example, within time $T > 0$ from the earth ($\boldsymbol{x}(0) = \boldsymbol{\xi}$) to the moon $\boldsymbol{x}(T) = \boldsymbol{0}$ with minimum fuel consumption. This is a typical problem of control.

If the control $u(t)$ for $t \geq 0$ depends only on the initial state $\boldsymbol{x}(0) = \boldsymbol{\xi}$, then the control is called *feedforward control*. Instead, if the control $u(t)$ for $t \geq 0$ is determined by a constant (or an intermittent) observation of the state $\boldsymbol{x}(\tau)$ with $0 \leq \tau \leq t$, then the control is called *feedback control*. Feedforward control uses only one observation $\boldsymbol{x}(0)$ at time $t = 0$. This is, so to speak, driving a bicycle (or a car) with eyes closed, while feedback control uses information from the eyes which is always (or sometimes) open. From this observation, we can easily understand that feedforward control is very fragile against uncertainties and disturbances. The feedback structure solves this fragility and leads to *robustness*. However, we mainly consider feedforward control since it gives clear mathematical structures of the optimal control. For feedback control implementation, one can adopt the receding horizon control, also known as the model predictive control [88] as discussed in Section 7.4.3, or self-triggered control [56].

8.1.2 Controllability and controllable set

We can consider many types of objectives of controlling the plant (8.1). For example, we set several target points $\boldsymbol{x}_1, \ldots, \boldsymbol{x}_s$ to control the plant so

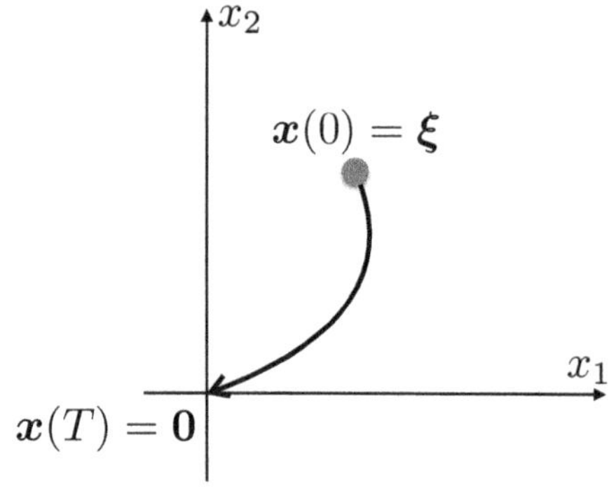

Figure 8.2: State transfer problem: finding a control $u(t)$, $0 \leq t \leq T$ that drives the state $\boldsymbol{x}(t)$ from a given initial state $\boldsymbol{x}(0) = \boldsymbol{\xi}$ to $\mathbf{0}$.

that the state $\boldsymbol{x}(t)$ passes approximately through these points at time $t = T_1, \ldots, T_s$, that is, $\boldsymbol{x}(T_i) \approx \boldsymbol{x}_i$. This control is called *trajectory generation*, or *trajectory planning*. We can also consider a control problem to keep the state $\boldsymbol{x}(t)$, $t \geq 0$, in a prescribed set $\mathcal{X}$ in the state space, that is, $\boldsymbol{x}(t) \in \mathcal{X}$ for all $t \geq 0$, assuming $\boldsymbol{x}(0) \in \mathcal{X}$. This problem arises, for example, in keeping a drone hovering in a region.

In this book, we mainly focus on the problem of *state transfer*. This problem is finding a control $u(t)$ that drives the state $\boldsymbol{x}(t)$ from a given initial state $\boldsymbol{\xi}$ to the origin $\mathbf{0}$ in a given time $T > 0$ (see also Figure 8.2).

First, we discuss the existence of the control. For this, we introduce the notion of controllability.

Definition 8.1 (Controllability). We call the system (8.1) is *controllable* if for any initial state $\boldsymbol{x}(0) = \boldsymbol{\xi} \in \mathbb{R}^d$, there exists a time $T > 0$ and a control $u(t)$, $0 \leq t \leq T$ such that the state $\boldsymbol{x}(t)$ in (8.1) is driven to the origin at time $t = T$, that is $\boldsymbol{x}(T) = \mathbf{0}$.

If the system is not controllable, then there exists an initial state that cannot be achieved to the origin with any $u(t)$ in finite time. The controllability is a fundamental requirement for control systems, and in this book we always assume that the system (8.1) is controllable.

Given a linear system, to check its controllability is an easy task. In fact, we have the following theorem for the controllability:

Theorem 8.1. The dynamical system (8.1) is controllable if and only if any of the following equivalent conditions are satisfied:

1. The following matrix called the *controllability matrix*

$$M \triangleq \begin{bmatrix} \boldsymbol{b} & A\boldsymbol{b} & A^2\boldsymbol{b} & \dots & A^{d-1}\boldsymbol{b} \end{bmatrix} \tag{8.7}$$

is non-singular.

2. The following matrix called the *controllability grammian*

$$G(T) \triangleq \int_0^T e^{At}\boldsymbol{b}\boldsymbol{b}^\top e^{A^\top t} dt \tag{8.8}$$

is non-singular for any $T > 0$.

3. For any $\lambda \in \mathbb{C}$,

$$\operatorname{rank} \begin{bmatrix} A - \lambda I & \boldsymbol{b} \end{bmatrix} = d. \tag{8.9}$$

4. For any left eigenvector $\boldsymbol{v}^\top$ of A,

$$\boldsymbol{v}^\top \boldsymbol{b} \neq 0. \tag{8.10}$$

From this theorem, to check the controllability of the dynamical system (8.1) is just to compute the determinant of the matrix M.

The controllability of the system (8.1) is completely determined by the matrix pair $(A, \boldsymbol{b})$. From this, we often say the pair $(A, \boldsymbol{b})$ *is controllable*, which means the system (8.1) is controllable.

Example 8.2. Let us consider the rocket model (8.5) and (8.6) in Example 8.1. The controllability matrix is given by

$$M = \begin{bmatrix} \boldsymbol{b} & A\boldsymbol{b} \end{bmatrix} = \begin{bmatrix} 0 & 1 \\ 1 & 0 \end{bmatrix}. \tag{8.11}$$

Since this matrix is non-singular, the system is controllable from Theorem 8.1. □

Note that if the dynamical system (8.1) is controllable, then for any initial state $\boldsymbol{\xi} \in \mathbb{R}^d$, any final state $\boldsymbol{\zeta} \in \mathbb{R}^d$, and any time $T > 0$, there exists a control $u(t)$, $0 \leq t \leq T$ that drives the state $\boldsymbol{x}(t)$ from $\boldsymbol{x}(0) = \boldsymbol{\xi}$ to $\boldsymbol{x}(T) = \boldsymbol{\zeta}$.

Exercise 8.2. Prove the above fact.

In general, the shorter the time $T > 0$ is, the larger the magnitude and the shorter the support of $u(t)$ should be. The shape of $u(t)$ may approach something like the Dirac delta function when T approaches zero. However, in real-world systems, actuators have limitations on their output, preventing them from generating arbitrarily large control signals. For example, can you imagine a vehicle capable of traveling at 1000 km/h? Hence, we assume the following limitation on $u(t)$:

$$|u(t)| \leq 1, \quad \forall t \in [0, T]. \tag{8.12}$$

We call a control that satisfies this constraint an *admissible control.* In (8.12), we assume the maximum magnitude is normalized to one, but if the maximum magnitude is $U_{\max} > 0$ and the limitation is represented by

$$|u(t)| \leq U_{\max}, \quad \forall t \in [0, T], \tag{8.13}$$

then we can redefine the vector $\boldsymbol{b}$ in the plant (8.1) as

$$\boldsymbol{b}' \triangleq \frac{\boldsymbol{b}}{U_{\max}}, \tag{8.14}$$

then the limitation is reduced to (8.13).

Under the constraint (8.12), there may be an initial state $\boldsymbol{\xi}$ that cannot be steered to the origin by any admissible control $u(t)$ that satisfies (8.12) within time $T > 0$ even if the system is controllable. To discuss this, we introduce the notion of the T-controllable set:

Definition 8.2 (T-Controllable Set). Fix $T > 0$. The set of initial states that can be steered to the origin by some admissible control $u(t)$, $0 \leq t \leq T$, is called the *T-controllable set.* We denote this set by $\mathcal{R}(T)$.

Exercise 8.3. Prove that $\mathcal{R}(T)$ can be represented by

$$\mathcal{R}(T) = \left\{ -\int_0^T e^{-At} \boldsymbol{b} u(t) dt : |u(t)| \leq 1, \ \forall t \in [0, T] \right\}. \tag{8.15}$$

For the T-controllable set, we have the following theorem:

Theorem 8.2. For any $T > 0$, the T-controllable set $\mathcal{R}(T)$ is a bounded, closed, and convex set. Also, if $T_1 < T_2$ then $\mathcal{R}(T_1) \subset$

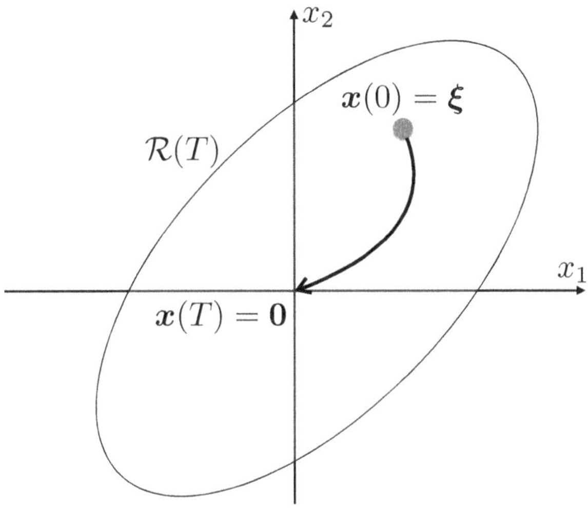

Figure 8.3: T-Controllable set $\mathcal{R}(T)$ in $\mathbb{R}^2$

$\mathcal{R}(T_2)$.

Exercise 8.4. Prove Theorem 8.2.

Figure 8.3 shows an illustration of a T-controllable set $\mathcal{R}(T)$ in $\mathbb{R}^2$. If an initial state $\boldsymbol{x}(0) = \boldsymbol{\xi}$ is in the T-controllable set $\mathcal{R}(T)$, then there exists an admissible control $u(t)$, $0 \le t \le T$, that steers the state to $\boldsymbol{x}(T) = \mathbf{0}$ in time T. If an initial state is outside the set $\mathcal{R}(T)$, then such control does not exist. We show an easy example to illustrate this property.

Example 8.3. Let us consider the problem of controlling a ball on an inclined plane shown in Figure 8.4. Let $x(t)$ denote the position of the ball on the x axis parallel to the slope. The origin is set at the top of the slope. The control objective is to move the ball from the initial position $x(0) < 0$ to the origin within time $T > 0$, that is, $x(T) = 0$.

The differential equation of $x(t)$ is given from Newton's second law of motion:

$$m\ddot{x}(t) = F(t) - mg\sin\theta. \tag{8.16}$$

Now, we assume

$$x(0) = -\xi, \quad \dot{x}(0) = 0, \tag{8.17}$$

where $\xi > 0$. From (8.16), we have

$$\dot{x}(t) = \frac{1}{m}\int_0^t F(\tau)d\tau - gt\sin\theta + \dot{x}(0), \tag{8.18}$$

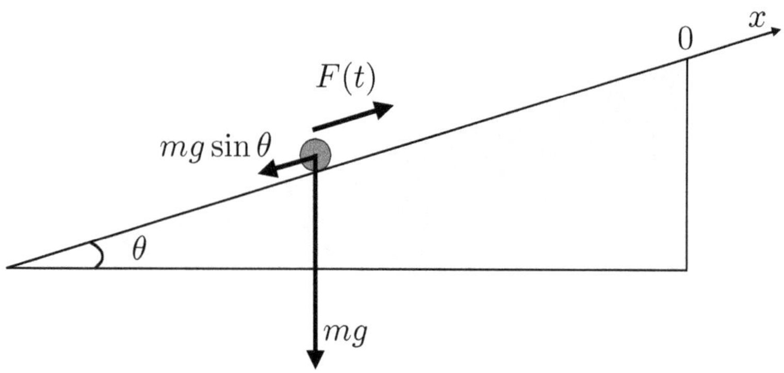

Figure 8.4: A ball on an inclined plane: m is the mass, g is the acceleration of gravity, θ is the angle of the slope, and $F(t)$ is the force (i.e., control) applied to the ball.

and

$$x(t) = \frac{1}{m}\int_0^t \int_0^s F(\tau)d\tau ds - \frac{1}{2}gt^2 \sin\theta + \dot{x}(0)t + x(0). \tag{8.19}$$

Suppose that there exists an admissible control $\{F(t) : 0 \le t \le T\}$ that satisfies

$$\|F\|_\infty \triangleq \sup_{t\in[0,T]} |F(t)| \le 1, \tag{8.20}$$

where $\|F\|_\infty$ is the L^∞ norm, such that $x(T) = 0$. Then we have

$$0 = x(T) = \frac{1}{m}\int_0^T \int_0^s F(\tau)d\tau ds - \frac{1}{2}gT^2 \sin\theta - \xi, \tag{8.21}$$

where we used the initial conditions in (8.17). From the above equation, we have

$$\begin{aligned} \left|\frac{1}{2}gT^2\sin\theta + \xi\right| &\le \frac{1}{m}\int_0^T\int_0^s |F(\tau)|d\tau ds \\ &\le \frac{1}{m}\int_0^T\int_0^s \|F\|_\infty d\tau ds \\ &= \frac{T^2}{2m}\|F\|_\infty. \end{aligned} \tag{8.22}$$

Since the variables m, g, T, and ξ are all positive, and $\sin\theta$ is also positive, we have

$$\|F\|_\infty \ge mg\sin\theta + \frac{2m\xi}{T^2}. \tag{8.23}$$

On the other hand, since F is an admissible control, it should satisfy

$$\|F\|_\infty \le 1. \tag{8.24}$$

From (8.23) and (8.24), we have

$$1 \geq mg\sin\theta + \frac{2m\xi}{T^2}. \tag{8.25}$$

It follows that $mg\sin\theta < 1$ and

$$T \geq \sqrt{\frac{2m\xi}{1 - mg\sin\theta}} \triangleq T^*. \tag{8.26}$$

From this, if the final time T is small such that $T < T^*$, then there is no admissible control with time T that achieves $x(T) = 0$. Conversely, let $T = T^*$ and take $F(t) \equiv 1$, $0 \leq t \leq T^*$. Then from (8.19), we can easily compute that

$$x(T^*) = 0. \tag{8.27}$$

Hence, $F(t) \equiv 1$ is a solution with $T = T^*$. Also, if $T > T^*$, we can choose $F(t)$ as

$$F(t) = \begin{cases} 1, & \text{if } \ 0 \leq t \leq T^*, \\ mg\sin\theta, & \text{if } \ T^* < t \leq T, \end{cases} \tag{8.28}$$

which is a solution with time T. □

From this example, the time T^* is the threshold that determines the T-controllability with a fixed initial state. The time T^* is called the *minimum time*, which is in general defined by

$$T^*(\boldsymbol{\xi}) \triangleq \inf\{T \geq 0 : \boldsymbol{\xi} \in \mathcal{R}(T)\}. \tag{8.29}$$

To consider the minimum time, we define the *controllable set* by the union of all $\mathcal{R}(T)$ with $T > 0$, that is,

$$\mathcal{R} \triangleq \bigcup_{T>0} \mathcal{R}(T). \tag{8.30}$$

Even if the system is controllable, the controllable set $\mathcal{R}$ may not be $\mathbb{R}^d$. That is, $\mathcal{R}$ may be a strict subset of $\mathbb{R}^d$. Then, if $\boldsymbol{\xi} \notin \mathcal{R}$, then there exists no admissible control on any finite time interval $[0, T]$ that achieves $\boldsymbol{x}(T) = \boldsymbol{0}$. For this case, we write $T^*(\boldsymbol{\xi}) = \infty$. Conversely, if $\boldsymbol{\xi} \in \mathcal{R}$ then the set $\{T \geq 0 : \boldsymbol{\xi} \in \mathcal{R}(T)\}$ is non-empty, and the minimum time (8.29) is finite, that is, $T^*(\boldsymbol{\xi}) < \infty$.

Assume that $\boldsymbol{\xi} \in \mathcal{R}$. Then $T^*(\boldsymbol{\xi}) < \infty$. Let us consider T_1 and T_2 such that

$$T_1 < T^*(\boldsymbol{\xi}) < T_2. \tag{8.31}$$

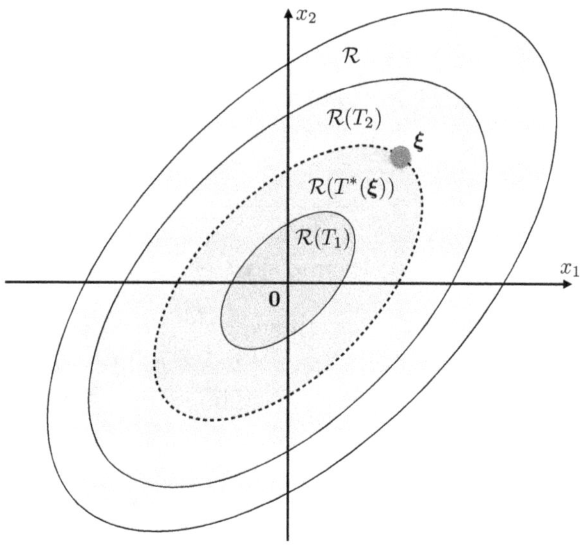

Figure 8.5: Controllable sets $\mathcal{R}(T_1) \subset \mathcal{R}(T^*(\boldsymbol{\xi})) \subset \mathcal{R}(T_2) \subset \mathcal{R}$, where $T_1 < T^*(\boldsymbol{\xi}) < T_2$. The minimum time $T^*(\boldsymbol{\xi})$ is the threshold for the feasibility from the initial state $\boldsymbol{\xi}$.

Then we have

$$\mathcal{R}(T_1) \subset \mathcal{R}(T^*(\boldsymbol{\xi})) \subset \mathcal{R}(T_2) \subset \mathcal{R}. \tag{8.32}$$

This inclusion relation is shown in Figure 8.5. From this figure, we have $\boldsymbol{\xi} \in \mathcal{R}(T_2)$ and $\boldsymbol{\xi} \notin \mathcal{R}(T_1)$. This means that if the final time is greater than $T^*(\boldsymbol{\xi})$, then there exists a feasible control, while if it is less than $T^*(\boldsymbol{\xi})$, then there is no control. The minimum time $T^*(\boldsymbol{\xi})$ is the threshold for the controllability, that is, $\boldsymbol{\xi}$ is on the boundary of the $T^*(\boldsymbol{\xi})$-controllability set $\mathcal{R}(T^*(\boldsymbol{\xi}))$. We will discuss the minimum time further in the next subsection.

For a stable system, the minimum time always exists for any initial state. In fact, we have the following theorem [57, Theorem 17.6]:

Theorem 8.3. Assume that $(A, \boldsymbol{b})$ is controllable. Assume also that A is stable, that is,

$$\lambda(A) \subset \mathbb{C}_- \triangleq \{z \in \mathbb{C} : \text{Re } z \leq 0\}, \tag{8.33}$$

where $\lambda(A)$ is the set of eigenvalues of A. Then the controllable set $\mathcal{R}$ is $\mathbb{R}^d$, and the minimum time $T^*(\boldsymbol{\xi})$ is finite for any $\boldsymbol{\xi} \in \mathbb{R}^d$.

8.1.3 Feasible control and minimum-time control

Fix $T > 0$ and assume $\boldsymbol{x}(0) = \boldsymbol{\xi} \in \mathcal{R}(T)$. Then by the definition of the T-controllable set in Definition 8.2, there exists an admissible control $u(t)$ that steers the state from $\boldsymbol{x}(0)$ to $\boldsymbol{x}(T) = \mathbf{0}$. We call such a control a *feasible control*. Let $\mathcal{U}(T, \boldsymbol{\xi})$ denote the set of feasible controls with initial state $\boldsymbol{\xi}$ and final time T. This set can be represented by

$$\mathcal{U}(T, \boldsymbol{\xi}) = \left\{ u \in L^\infty(0, T) : \boldsymbol{\xi} = -\int_0^T e^{-At} \boldsymbol{b} u(t) dt, \ |u(t)| \leq 1, \ \forall t \in [0, T] \right\}. \tag{8.34}$$

Exercise 8.5. Prove that the set of feasible controls is represented by (8.34).

It is easily shown that $\boldsymbol{\xi} \in \mathcal{R}(T)$ if and only if there exists an admissible control u such that $u \in \mathcal{U}(T, \boldsymbol{\xi})$. Hence the minimum time $T^*(\boldsymbol{\xi})$ in (8.29) is formulated by

$$T^*(\boldsymbol{\xi}) = \inf\{T \geq 0 : \exists u, \ u \in \mathcal{U}(T, \boldsymbol{\xi})\}. \tag{8.35}$$

From the discussion in the previous subsection, $T^*(\boldsymbol{\xi})$ is finite if and only if $\boldsymbol{\xi} \in \mathcal{R}$. From this, if $\boldsymbol{\xi} \in \mathcal{R}$, then there exists a final time $T \geq 0$ and an admissible control u such that $u \in \mathcal{U}(T, \boldsymbol{\xi})$, and hence the set of the right hand side of (8.35) is non-empty.

Now we find the control that achieves the minimum time. That is, we consider the following optimization problem:

$$\underset{u}{\text{minimize}} \ T \quad \text{subject to} \quad u \in \mathcal{U}(T, \boldsymbol{\xi}). \tag{8.36}$$

The solution is called the *minimum-time control* or the *time-optimal control*. The minimum-time control exists if $\boldsymbol{\xi} \in \mathcal{R}$ or equivalently $T^*(\boldsymbol{\xi}) < \infty$. In fact, we have the following lemma:

Theorem 8.4. Assume $T^*(\boldsymbol{\xi}) < \infty$. Then there exists a minimum-time control $u^* \in \mathcal{U}(T^*(\boldsymbol{\xi}), \boldsymbol{\xi})$. Moreover, for any $T > T^*(\boldsymbol{\xi})$, $\mathcal{U}(T, \boldsymbol{\xi})$ is non-empty.

Exercise 8.6. Prove Theorem 8.4

8.1.4 Optimal control and Pontryagin minimum principle

From Theorem 8.4, if $\boldsymbol{\xi} \in \mathcal{R}$ and $T > T^*(\boldsymbol{\xi})$, then the set of feasible controls $\mathcal{U}(T, \boldsymbol{\xi})$ is non-empty, and in general $\mathcal{U}(T, \boldsymbol{\xi})$ has infinitely many elements.

Optimal control is the control that minimizes a given cost function among all feasible controls in $\mathcal{U}(T, \boldsymbol{\xi})$.

The following is the formulation of the optimal control that is mainly considered in Part II of this book:

Problem 8.1 (Optimal Control Problem). For the plant modeled by

$$\dot{\boldsymbol{x}}(t) = A\boldsymbol{x}(t) + \boldsymbol{b}u(t), \quad t \geq 0, \quad \boldsymbol{x}(0) = \boldsymbol{\xi} \in \mathbb{R}^d, \tag{8.37}$$

find an admissible control u (i.e., $\|u\|_\infty \leq 1$) that achieves

$$\boldsymbol{x}(T) = \boldsymbol{0}, \tag{8.38}$$

and minimizes the following cost function:

$$J(u) = \int_0^T L(u(t))dt. \tag{8.39}$$

We here assume that the function $L(u)$, called the *stage cost function*, is continuous in u. We call the solution the *optimal control*. Note that the optimal control problem can also be written by using the set of feasible controls $\mathcal{U}(T, \boldsymbol{\xi})$ as

$$\underset{u}{\text{minimize}} \ J(u) \ \text{ subject to } \ u \in \mathcal{U}(T, \boldsymbol{\xi}). \tag{8.40}$$

The minimum-time control (8.36) is the optimal control with $L(u) = 1$.

Let us assume that there exists an optimal control for Problem 8.1. We here introduce *Pontryagin's minimum principle* that gives necessary conditions for the optimal control.

First, define the following function called *Hamiltonian*:

$$H^\eta(\boldsymbol{x}, \boldsymbol{p}, u) \triangleq \boldsymbol{p}^\top(A\boldsymbol{x} + \boldsymbol{b}u) + \eta L(u), \tag{8.41}$$

where $\eta \in \{0, 1\}$ is called the *abnormal multiplier*. The following theorem is Pontryagin's minimum principle.

Theorem 8.5 (Pontryagin's Minimum Principle (PMP)). Assume that an optimal control u^* of the optimal control problem (Problem 8.1) exists. Let us denote by $\{\boldsymbol{x}^*(t) : 0 \leq t \leq T\}$ the *optimal state* with the optimal control $\{u^*(t) : 0 \leq t \leq T\}$, that is,

$$\boldsymbol{x}^*(t) \triangleq e^{At}\boldsymbol{\xi} + \int_0^t e^{A(t-\tau)}\boldsymbol{b}u^*(\tau)d\tau, \quad \forall t \in [0, T]. \tag{8.42}$$

Then there exist $\eta \in \{0, 1\}$ and the *optimal costate* $\{\boldsymbol{p}^*(t) : 0 \leq t \leq T\}$ that satisfy the following conditions.

(non-triviality condition) The abnormal multiplier η and the optimal costate $\boldsymbol{p}^*$ satisfy the *non-triviality condition*:

$$|\eta| + \|\boldsymbol{p}^*\|_\infty > 0. \tag{8.43}$$

(canonical equation) The following *canonical equations* hold:

$$\begin{aligned} \dot{\boldsymbol{x}}^*(t) &= A\boldsymbol{x}^*(t) + \boldsymbol{b}u^*(t), \\ \dot{\boldsymbol{p}}^*(t) &= -A^\top \boldsymbol{p}^*(t), \quad \forall t \in [0, T]. \end{aligned} \tag{8.44}$$

The differential equation for $\boldsymbol{p}^*(t)$ is called the *adjoint equation.*

(minimum condition) The optimal control $u^*(t)$ minimizes Hamiltonian at each time $t \in [0, T]$. That is,

$$u^*(t) = \underset{u \in [-1,1]}{\arg\min}\, H^\eta(\boldsymbol{x}^*(t), \boldsymbol{p}^*(t), u), \quad \forall t \in [0, T]. \tag{8.45}$$

(consistency) Hamiltonian satisfies

$$H^\eta(\boldsymbol{x}^*(t), \boldsymbol{p}^*(t), u^*(t)) = c, \quad \forall t \in [0, T], \tag{8.46}$$

where c is a constant independent of t. If T is not fixed (as in the minimum-time control), then

$$H^\eta(\boldsymbol{x}^*(t), \boldsymbol{p}^*(t), u^*(t)) = 0, \quad \forall t \in [0, T]. \tag{8.47}$$

Note that the canonical equation in (8.44) can be rewritten in terms of Hamiltonian H^η as

$$\begin{aligned} \dot{\boldsymbol{x}}^*(t) &= \frac{\partial H^\eta}{\partial \boldsymbol{p}}(\boldsymbol{x}^*(t), \boldsymbol{p}^*(t), u^*(t)), \\ \dot{\boldsymbol{p}}^*(t) &= -\frac{\partial H^\eta}{\partial \boldsymbol{x}}(\boldsymbol{x}^*(t), \boldsymbol{p}^*(t), u^*(t)), \quad \forall t \in [0, T]. \end{aligned} \tag{8.48}$$

These equations are also called *Hamilton's canonical equations*.

Pontryagin's minimum principle is a powerful tool to analyze the optimal control (if it exists). For simple problems, we can obtain a closed form of the control that satisfies the necessary conditions. We call this an *extremal control.* We should note that an extremal control is not necessarily the optimal control. However, in some cases, we can determine the optimal control from the minimum principle. One example is shown in Section 8.3, the minimum-time control for the rocket in Example 8.1. Before the example, we will formulate the minimum-time control for general linear systems.

8.2 Minimum-time Control

Let us consider the following linear system:

$$\dot{\boldsymbol{x}}(t) = A\boldsymbol{x}(t) + \boldsymbol{b}u(t), \quad t \geq 0, \quad \boldsymbol{x}(0) = \boldsymbol{\xi} \in \mathbb{R}^d. \tag{8.49}$$

For this system, we consider the minimum-time control, which is given by the optimal control problem (Problem 8.1) with the stage cost $L(u) = 1$. Then the Hamiltonian is given by

$$H^\eta(\boldsymbol{x}, \boldsymbol{p}, u) = \boldsymbol{p}^\top(A\boldsymbol{x} + \boldsymbol{b}u) + \eta. \tag{8.50}$$

Let us assume that the minimum-time control exists. Then from Pontryagin's minimum principle, the optimal control $u^*(t)$ should satisfy

$$u^*(t) = \arg\min_{u\in[-1,1]} H^\eta(\boldsymbol{x}^*(t), \boldsymbol{p}^*(t), u), \quad \forall t \in [0, T^*(\boldsymbol{\xi})], \tag{8.51}$$

where $\boldsymbol{x}^*(t)$ and $\boldsymbol{p}^*(t)$ are respectively the optimal state and costate by the optimal control $u^*(t)$, and $T^*(\boldsymbol{\xi})$ is the minimum time. From this, we have

$$u^*(t) = \arg\min_{u\in[-1,1]} \boldsymbol{p}^*(t)^\top \boldsymbol{b}u = -\mathrm{sgn}(\boldsymbol{p}^*(t)^\top \boldsymbol{b}), \tag{8.52}$$

where $\mathrm{sgn}(\cdot)$ is the *sign function* defined by

$$\begin{aligned} \mathrm{sgn}(a) &= \begin{cases} 1, & a > 0, \\ -1, & a < 0, \end{cases} \\ \mathrm{sgn}(a) &\in [-1, 1], \quad a = 0. \end{aligned} \tag{8.53}$$

If the function $\boldsymbol{p}^*(t)^\top \boldsymbol{b}$ is not zero for almost all $t \in [0, T^*(\boldsymbol{\xi})]$, then the control $u^*(t)$ takes values of only ± 1 for almost all t. Such a control is called a *bang-bang control.*

Lemma 8.1. If $(A, \boldsymbol{b})$ is controllable, then the function $\boldsymbol{p}^*(t)^\top \boldsymbol{b}$ is not zero for almost all $t \in [0, T^*(\boldsymbol{\xi})]$.

Exercise 8.7. Prove Lemma 8.1.

For the minimum-time control problem, we have the following existence and uniqueness theorems.

Theorem 8.6 (Existence). If the initial state $\boldsymbol{\xi}$ is in the controllable set $\mathcal{R}$ defined in (8.30), then a minimum-time control exists.

Theorem 8.7 (Uniqueness). Assume that $(A, \boldsymbol{b})$ is controllable. Then the minimum-time control is (if it exists) unique.

Exercise 8.8. Prove Theorems 8.6 and 8.7.

The following corollary is easily proved from Theorems 8.3, 8.6, and 8.7.

Corollary 8.1. Assume that $(A, \boldsymbol{b})$ is controllable and A is stable. Then for any $\boldsymbol{\xi} \in \mathbb{R}^d$, the minimum-time control $u^* \in \mathcal{U}(\boldsymbol{\xi})$ uniquely exists.

8.3 Rocket Control Example

Here we derive the minimum-time control of the rocket in Example 8.1 by using Pontryagin's minimum principle.

Now, from Example 8.2, the pair $(A, \boldsymbol{b})$ is controllable. It is also easily seen that A is stable since A has a multiple eigenvalue of 0. Therefore, from Theorem 8.1, there uniquely exists the minimum-time control u^*.

Let us define the optimal state and costate by

$$\boldsymbol{x}^*(t) = \begin{bmatrix} x_1^*(t) \\ x_2^*(t) \end{bmatrix}, \quad \boldsymbol{p}^*(t) = \begin{bmatrix} p_1^*(t) \\ p_2^*(t) \end{bmatrix}. \tag{8.54}$$

For simplicity, we assume the mass of the rocket $m = 1$.

Then the Hamiltonian $H^\eta(\boldsymbol{x}, \boldsymbol{p}, u)$ in (8.50) is given by

$$H^\eta(\boldsymbol{x}, \boldsymbol{p}, u) = \boldsymbol{p}^\top (A\boldsymbol{x} + \boldsymbol{b}u) + \eta = p_1 x_2 + p_2 u + \eta. \tag{8.55}$$

The canonical equation (8.44) for the costate $\boldsymbol{p}^*(t)$ is given by

$$\begin{aligned} \dot{p}_1^*(t) &= 0, \\ \dot{p}_2^*(t) &= -p_1^*(t). \end{aligned} \tag{8.56}$$

Let $p_1^*(0) = \pi_1$ and $p_2^*(0) = \pi_2$. Then the solution to the differential equation (8.56) is given by

$$\begin{aligned} p_1^*(t) &= \pi_1, \\ p_2^*(t) &= \pi_2 - \pi_1 t. \end{aligned} \tag{8.57}$$

Since T is not fixed, from the condition (8.47), we have $H^\eta(\boldsymbol{x}^*(t), \boldsymbol{p}^*(t), u^*) = 0$. That is,

$$p_1^*(t)x_2^*(t) + p_2^*(t)u^* + \eta = 0. \tag{8.58}$$

If $\pi_1 = \pi_2 = 0$, then $p_1^*(t) = p_2^*(t) = 0$ from (8.57), and hence $\eta = 0$ from (8.58). But this contradicts the non-triviality condition (8.43). Therefore, $\pi_1 \neq 0$ or $\pi_2 \neq 0$, that is, $\boldsymbol{p}^*(0) \neq \boldsymbol{0}$.

Next, from (8.52), we have

$$u^*(t) = -\text{sgn}(\boldsymbol{p}^*(t)^\top \boldsymbol{b}) = -\text{sgn}(p_2^*(t)). \tag{8.59}$$

From (8.57), $p_2^*(t)$ is a linear function $p_2^*(t) = \pi_2 - \pi_1 t$. Then we need to check the following cases:

(i) $\pi_1 \leq 0, \pi_2 \leq 0$ with $(\pi_1, \pi_2) \neq (0,0)$,

(ii) $\pi_1 \geq 0, \pi_2 \geq 0$ with $(\pi_1, \pi_2) \neq (0,0)$,

(iii) $\pi_1 < 0$, $\pi_2 > 0$,

(iv) $\pi_1 > 0$, $\pi_2 < 0$.

Extremum controls given in (8.59) for the four cases are shown in Figure 8.6. From this figure, it is easily shown that each extremum control takes its values of ± 1 for almost all t, that is, bang-bang. We note that the number of switching is at most one from Figure 8.6.

Next, we compute the trajectory $\boldsymbol{x}(t)$ when $u(t)$ is constant (i.e., ± 1). From (8.5), we have

$$\begin{aligned} \dot{x}_1(t) &= x_2(t), \quad x_1(0) = \xi_1, \\ \dot{x}_2(t) &= u(t), \quad x_2(0) = \xi_2. \end{aligned} \tag{8.60}$$

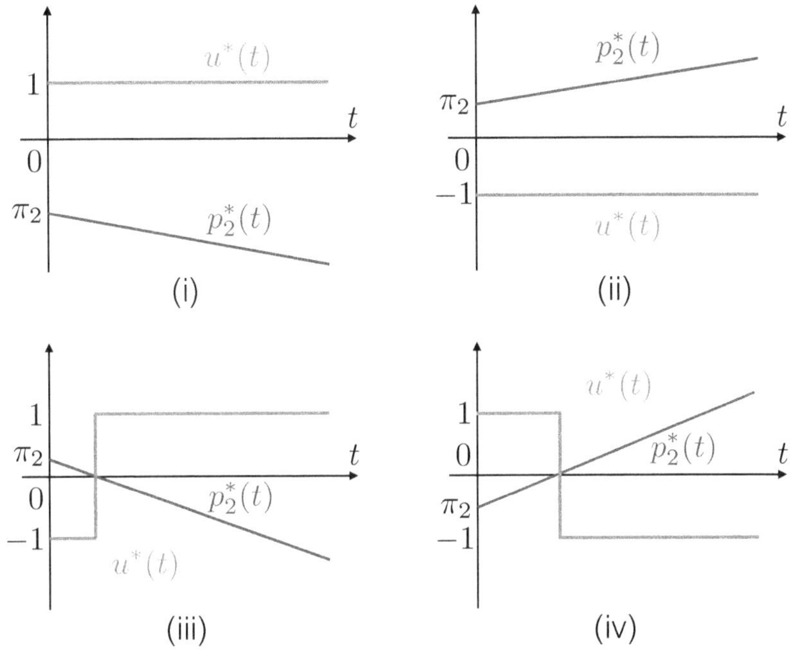

Figure 8.6: Costate $p_2^*(t) = \pi_2 - \pi_1 t$ and corresponding extremum control $u^*(t)$ from (8.59)

If $u(t) = \pm 1$, then

$$\begin{aligned} x_1(t) &= \pm\frac{1}{2}t^2 + \xi_2 t + \xi_1, \\ x_2(t) &= \pm t + \xi_2. \end{aligned} \tag{8.61}$$

Eliminating the time variable t gives

$$x_1(t) = \pm\frac{1}{2}x_2(t)^2 + \xi_1 \mp \frac{1}{2}\xi_2^2. \tag{8.62}$$

That is, when the control $u(t)$ is constant ± 1, then the state $(x_1(t), x_2(t))$ moves on the following parabolic curves:

$$x_1 = \frac{1}{2}x_2^2 + \xi_1 - \frac{1}{2}\xi_2^2, \quad \text{if } u(t) = 1, \tag{8.63}$$

$$x_1 = -\frac{1}{2}x_2^2 + \xi_1 + \frac{1}{2}\xi_2^2, \quad \text{if } u(t) = -1. \tag{8.64}$$

Figure 8.7 shows the flow of the state $(x_1(t), x_2(t))$ by some of these parabolic curves with directions of the state to move. Note that the parabolic curves defined in (8.63) and (8.64) go through the point (ξ_1, ξ_2).

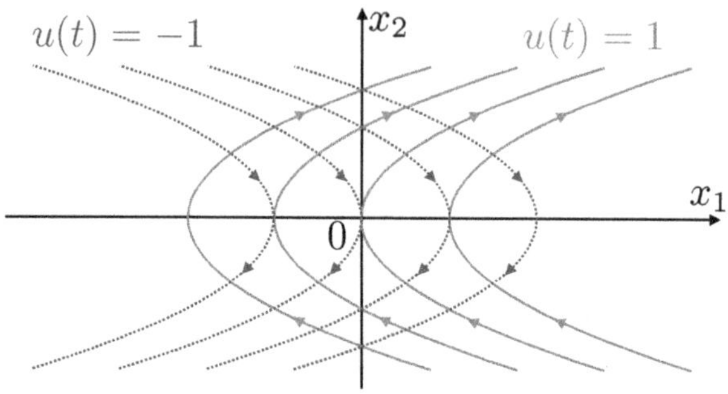

Figure 8.7: Flow of state $(x_1(t), x_2(t))$ by constant control $u(t) = 1$ (solid curve) and $u(t) = -1$ (dashed curve).

To achieve the terminal state $\boldsymbol{x}(T) = \boldsymbol{0}$, the final trajectory must be on the parabolic curve that goes through the origin:

$$\begin{aligned} x_1 &= \frac{1}{2}x_2^2, \qquad \text{if } u(t) = 1, \\ x_1 &= -\frac{1}{2}x_2^2, \qquad \text{if } u(t) = -1. \end{aligned} \tag{8.65}$$

From this, if there is no switching, that is, in the cases of (see Figure 8.7)

(i) $u^*(t) \equiv 1$,

(ii) $u^*(t) \equiv -1$,

then, the initial state (ξ_1, ξ_2) should be on the parabolic curve

$$\gamma_+ \triangleq \{(x_1, x_2) \in \mathbb{R}^2 : x_1 = x_2^2/2, \quad x_2 \leq 0\}, \tag{8.66}$$

or

$$\gamma_- \triangleq \{(x_1, x_2) \in \mathbb{R}^2 : x_1 = -x_2^2/2, \quad x_2 \leq 0\}. \tag{8.67}$$

Figure 8.8 shows the two curves γ_+ and γ_-. In fact, we can easily show that

- if the initial state (ξ_1, ξ_2) is on the curve γ_+, then $u^*(t) \equiv 1$ is the unique extremum control.
- if the initial state (ξ_1, ξ_2) is on the curve γ_-, then $u^*(t) \equiv -1$ is the unique extremum control.

The proof is shown below.

Assume $(\xi_1, \xi_2) \in \gamma_+$. As mentioned above, there are four extremum controls with (i)–(iv). Now, $u^*(t) \equiv 1$ is for the case (i). The point A

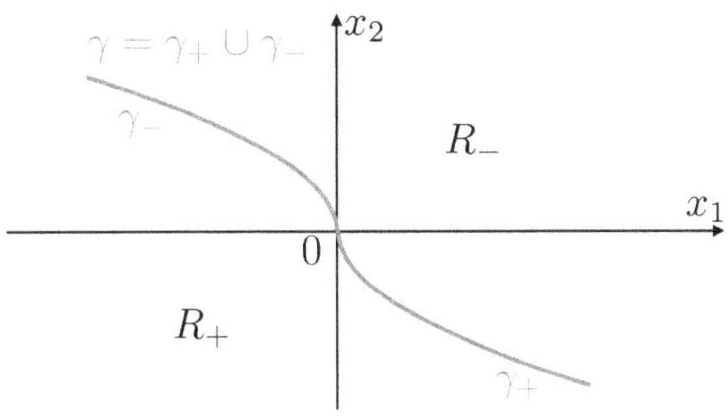

Figure 8.8: Switching curve $\gamma = \gamma_+ \cup \gamma_-$ and regions R_+ and R_-

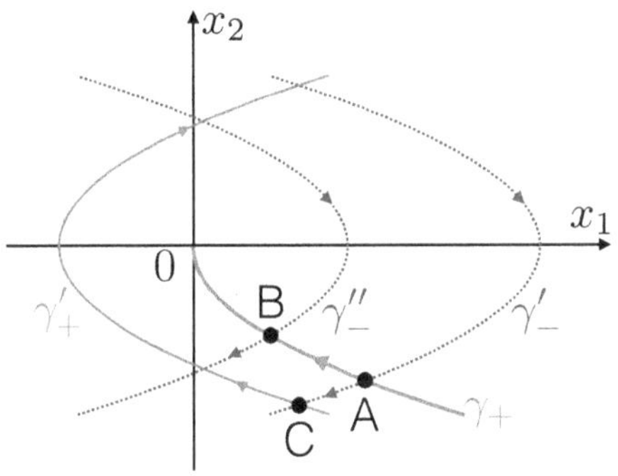

Figure 8.9: Four cases of state trajectories from initial state (ξ_1, ξ_2) on the curve γ_+.

in Figure 8.9 is the initial point, and the state can reach the origin by $u^*(t) \equiv 1$ through the curve γ_+. However, for the other cases (ii), (iii), and (iv), the state never reaches the origin from the initial point A. For the case (ii), by the control $u^*(t) \equiv -1$, the state starts at A on the curve γ'_- to the direction to C, and never reaches the origin. For the case (iii), the state moves on the curve γ'_- from A to C by the control $u^*(t) = -1$, which is switched to $u^*(t) = +1$ at C. Then the state moves on the curve γ'_+, which never reaches the origin. Finally, for the case (iv), the state starts from A to B on the curve γ'_+ by the control $u^*(t) = +1$, which is switched to $u^*(t) = -1$ at B. The state then moves from B on the curve γ''_- to the indicated direction and never reaches the origin. In summary, (i) $u^*(t) \equiv 1$ is the unique extremum control, and hence if the minimum-time control exists, this is actually the optimal control. The same discussion can be applied for the initial state (ξ_1, ξ_2) on the curve γ_-, and the unique extremum control is $u^*(t) \equiv -1$.

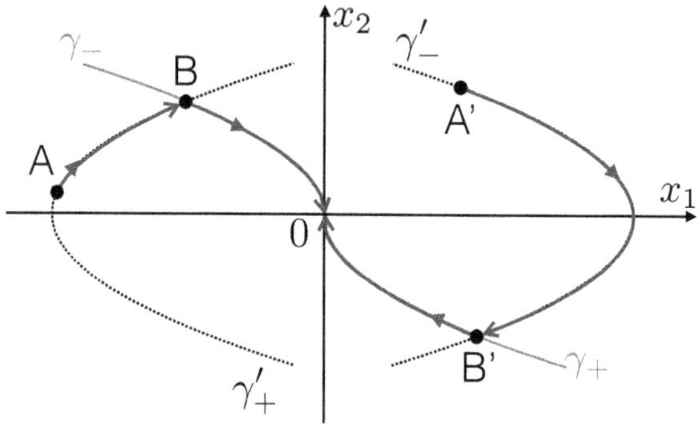

Figure 8.10: State trajectories from initial points A and A'

Next, let us consider the initial state (ξ_1, ξ_2) is outside the curve

$$\gamma \triangleq \gamma_+ \cup \gamma_- = \{(x_1, x_2) \in \mathbb{R}^2 : x_1 = -x_2|x_2|/2\}. \tag{8.68}$$

Let us define two regions R_+ and R_- divided by the curve γ:

$$\begin{aligned} R_+ &\triangleq \{(x_1, x_2) \in \mathbb{R}^2 : x_1 < -x_2|x_2|/2\}, \\ R_- &\triangleq \{(x_1, x_2) \in \mathbb{R}^2 : x_1 > -x_2|x_2|/2\}. \end{aligned} \tag{8.69}$$

Figure 8.8 shows these regions. We call the curve γ the *switching curve.*

Now assume the initial state (ξ_1, ξ_2) is at A in R_+ as in Figure 8.10. From the point A, the curve γ'_+ defined in (8.63) is plotted. By a constant control $u(t) \equiv 1$, the state moves on the curve γ'_+ to the indicated direction from A. At some time, the state touches the switching curve γ_- at B, and the control is switched to $u(t) = -1$. From B, the state goes on the switching curve to the origin. The control switched from $+1$ to -1 is the control in (iii) in Figure 8.6. We can easily show that this is the unique extremum control from any initial point in R_+, in a similar way to the case where the initial state is on the curve $\gamma = \gamma_+ \cup \gamma_-$.

Then, let us consider the initial state $(\xi_1, \xi_2) \in R_-$ at A' in Figure 8.10. First, by the constant control $u(t) = -1$, the state moves on the curve γ'_- from A' to B'. Then the control is switched from -1 to $+1$, and the state is steered to the origin along the curve γ_+. This control is for the case (iv) in Figure 8.6, and actually this is the unique extremum solution.

In summary, the extremum control $u^*(t)$ of the minimum-time control problem is given by

$$u^*(t) = \begin{cases} 1, & \text{if } \boldsymbol{x}(t) \in \gamma_+ \cup R_+ \setminus \{\boldsymbol{0}\}, \\ -1, & \text{if } \boldsymbol{x}(t) \in \gamma_- \cup R_- \setminus \{\boldsymbol{0}\}, \\ 0, & \text{if } \boldsymbol{x}(t) = \boldsymbol{0}. \end{cases} \tag{8.70}$$

The control $u^*(t)$ depends on the state $\boldsymbol{x}(t)$, and hence the control is a *feedback control*, which changes its value (± 1 or 0) based on the observation of the state $\boldsymbol{x}(t)$.

Exercise 8.9. Compute the minimum time $T^*(\boldsymbol{\xi})$ from $\boldsymbol{\xi} = (\xi_1, \xi_2)$ to the origin, and the switching time when $\boldsymbol{\xi} \in R_+$ and $\boldsymbol{\xi} \in R_-$.

8.4 Further Readings

For the basics of control theory with state-space formulations, I recommend a nice book by Ogata [119]. For the controllable set, see [138]. The proof of Pontryagin's minimum principle is found in [82]. Pontryagin's minimum principle is also referred to as Pontryagin's maximum principle, which is mathematically equivalent to the minimum principle. The book by Clarke [28] is one of the most reliable books on Pontryagin's maximum principle. For the minimum-time control, see the classical books of [3], [57], [128].

Chapter 9

Maximum Hands-off Control

In this chapter, we introduce a new optimal control problem called maximum hands-off control, which is the sparsest control among all feasible controls.

Key ideas of Chapter 9

- Maximum hands-off control is described as L^0-optimal control.
- Under the assumption of non-singularity, L^0-optimal control is equivalent to L^1-optimal control.
- Maximum hands-off control is a ternary signal that takes values of ± 1 and 0. Such a ternary control is called bang-off-bang control.

9.1 L^0 Norm and Sparsity

Here we introduce mathematical preliminaries for maximum hands-off control.

First, we define the *support* of a function $u(t)$ on a finite interval $[0, T]$ by

$$\operatorname{supp}(u) \triangleq \{t \in [0, T] : u(t) \neq 0\}. \tag{9.1}$$

By using the support, we define the L^0 *norm* by the length of the support of function u, that is,

$$\|u\|_0 \triangleq \mu(\operatorname{supp}(u)), \tag{9.2}$$

where $\mu(S)$ is the Lebesgue measure of a subset $S \subset [0, T]$. From this definition, the L^0 norm of a continuous-time signal is the total length of time duration on which the signal takes nonzero values.

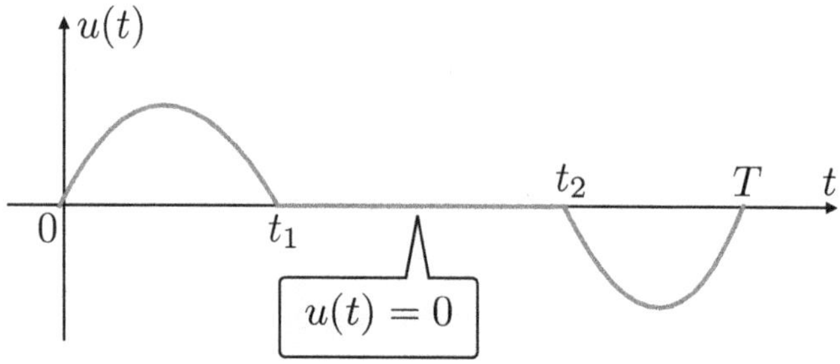

Figure 9.1: The L^0 norm of the function $u(t)$ is $t_1 + (T - t_2)$.

The L^0 norm can be represented as an integral. Define

$$|u|^0 \triangleq \begin{cases} 0, & \text{if } u = 0, \\ 1, & \text{if } u \neq 0, \end{cases} \tag{9.3}$$

then the L^0 norm in (9.2) can be written as

$$\|u\|_0 = \int_0^T |u(t)|^0 dt. \tag{9.4}$$

Example 9.1. Let us consider a function $u(t)$, as shown in Figure 9.1. The function $u(t)$ is zero over the interval $[t_1, t_2]$, and the support set of u is

$$\text{supp}(u) = (0, t_1) \cup (t_2, T) \subset [0, T]. \tag{9.5}$$

From this, the L^0 norm of u is

$$\|u\|_0 = \mu(\text{supp}(u)) = t_1 + (T - t_2) = T - (t_2 - t_1). \tag{9.6}$$

□

In the above example, the value $t_2 - t_1$ is the length of the interval $[t_1, t_2]$ on which $u(t) = 0$. If $\|u\|_0$ is much smaller than the total length T (i.e., $\|u\|_0 \ll T$), then the signal is said to be *sparse*. This notion is an analogy of the sparsity of vectors studied in Part I of this book.

Note that the L^0 norm does not have the absolute homogeneous property (see Definition 2.1, p. 21). In fact, if we take a non-zero scalar α such that $|\alpha| \neq 1$, then

$$\|\alpha u\|_0 = \|u\|_0 \neq |\alpha| \|u\|_0. \tag{9.7}$$

Note also that a sparse signal $u(t)$ on $[0, T]$ has a time duration whose length is positive, on which the control $u(t)$ is exactly zero. This means that the function $u(t)$ is not a real analytic function[1]. For example, a polynomial function

$$p(t) = t^n + a_{n-1}t^{n-1} + \cdots + a_1 t + a_0, \tag{9.8}$$

a trigonometric function $\sin(\omega t)$ $(\omega \neq 0)$, an exponential function $e^{\lambda t}$, and their sum or product are never sparse.

9.2 Practical Benefits of Sparsity in Control

Let us consider a sparse control signal $u(t)$, $t \in [0, T]$, as depicted in Figure 9.1. This control signal is exactly zero on the time interval $[t_1, t_2]$. In an electromechanical system, the control signal is typically transformed into mechanical motion by an actuator, such as an electric motor. This transformation often involves an amplifier attached between the controller and the actuator to provide sufficient energy to drive the actuator and generate the desired mechanical motion. Consequently, effective actuation requires not only a suitable control signal but also an adequate energy supply.

By using a sparse signal as in Figure 9.1, we can effectively deactivate the actuator over the time interval $[t_1, t_2]$. This strategic deactivation allows for significant energy savings, such as reduced consumption of electric power or fuel during this period. This control strategy, characterized by periods of actuator inactivity, is referred to as *hands-off control*, also known as *gliding* or *coasting*.

This control strategy is actually used in practical control systems. A stop-start system [40], [77] in automobiles is an example of hands-off control. It automatically shuts down the engine to avoid it idling for a long duration of time. By this, we can reduce CO or CO2 emissions as well as fuel consumption. Also in hybrid vehicles [21], [115], [140], the internal combustion engine is stopped when the vehicle is at a stop or the speed is lower than a preset threshold, and the electric motor is alternatively used. Other examples are found in railway vehicles [22], [76] and free-flying robots [150].

Hands-off control is also desirable for networked and embedded systems. During periods of zero-valued control, communication can be temporarily

[1]A function $u(t)$ is said to be *real analytic* if it is an infinitely differentiable function such that the Taylor series at any point $t_0 \in (0, T)$ converges to $u(t)$ for t in a neighborhood of t_0 pointwise. See [132, Chapter 8] for details.

suspended, which is advantageous in particular for wireless communications [71].

Due to the characteristics mentioned above, hands-off control is also known as *green control.*

9.3 Problem Formulation of Maximum Hands-off Control

Let us consider the optimal control problem (Problem 8.1, p. 186) with the stage cost function

$$L(u) = |u|^0. \tag{9.9}$$

This is called an L^0*-optimal control problem* or a *maximum hands-off control problem.*

Problem 9.1 (L^0-optimal control problem). For the linear time-invariant system

$$\dot{\boldsymbol{x}}(t) = A\boldsymbol{x}(t) + \boldsymbol{b}u(t), \quad t \geq 0, \quad \boldsymbol{x}(0) = \boldsymbol{\xi} \in \mathbb{R}^d, \tag{9.10}$$

find a control $\{u(t) : t \in [0, T]\}$ with $T > 0$ that minimizes

$$J_0(u) = \|u\|_0 = \int_0^T |u(t)|^0 dt, \tag{9.11}$$

subject to

$$\boldsymbol{x}(T) = \boldsymbol{0}, \tag{9.12}$$

and

$$\|u\|_\infty \leq 1. \tag{9.13}$$

The solution of this optimal control problem is called the L^0*-optimal control*, or the *maximum hands-off control.*

The stage cost function (9.9) is discontinuous and non-convex, as shown in Figure 9.2. By borrowing the idea of sparse representation to use the ℓ^1 norm for ℓ^0 norm optimization, we introduce the following cost function with the L^1 norm:

$$J_1(u) \triangleq \|u\|_1 = \int_0^T |u(t)| dt. \tag{9.14}$$

As shown in Figure 9.2, the stage cost function $|u|$ is an approximation of $|u|^0$. In fact, this approximation is mathematically explained as the *convex relaxation.* That is, the L^1 norm $\|u\|_1$ is the convex relaxation of $\|u\|_0$ when $\|u\|_\infty \leq 1$. See [149, Section 1.3.2] for details.

Now we formulate the L^1*-optimal control problem.*

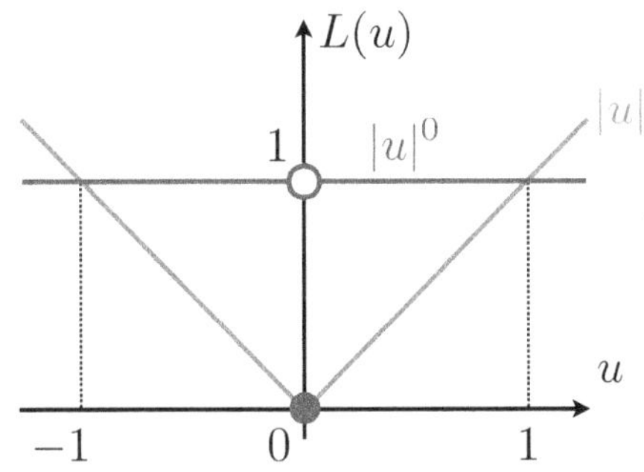

Figure 9.2: Stage cost functions $|u|^0$ and $|u|$ in L^0 and L^1 optimal control problems.

Problem 9.2 (L^1-optimal control problem). For the linear time-invariant system

$$\dot{\boldsymbol{x}}(t) = A\boldsymbol{x}(t) + \boldsymbol{b}u(t), \quad t \geq 0, \quad \boldsymbol{x}(0) = \boldsymbol{\xi} \in \mathbb{R}^d, \tag{9.15}$$

find a control $\{u(t) : t \in [0, T]\}$ with $T > 0$ that minimizes

$$J_1(u) = \|u\|_1 = \int_0^T |u(t)|dt, \tag{9.16}$$

subject to

$$\boldsymbol{x}(T) = \boldsymbol{0}, \tag{9.17}$$

and

$$\|u\|_\infty \leq 1. \tag{9.18}$$

We call the solution of this optimal control problem the L^1*-optimal control.* This optimal control is also known as *minimum-fuel control*, which was widely studied in the 60s for rocket control.

The L^1-optimal control (Problem 9.2) is a convex optimization problem since the stage cost $L(u) = |u|$ is convex in u and the constraints are also convex. Although the variable u is a function, which is a member of the infinite dimensional function space $L^\infty(0, T)$, the problem can be reduced to a finite-dimensional optimization problem via time discretization. See Chapter 10 for details.

9.4 L^1-optimal Control

Here we investigate properties of L^1 optimal control by using necessary conditions from Pontryagin's minimum principle.

For the L^1-optimal control problem (Problem 9.2), the Hamiltonian is given by

$$H^\eta(\boldsymbol{x}, \boldsymbol{p}, u) = \boldsymbol{p}^\top(A\boldsymbol{x} + \boldsymbol{b}u) + \eta|u|. \tag{9.19}$$

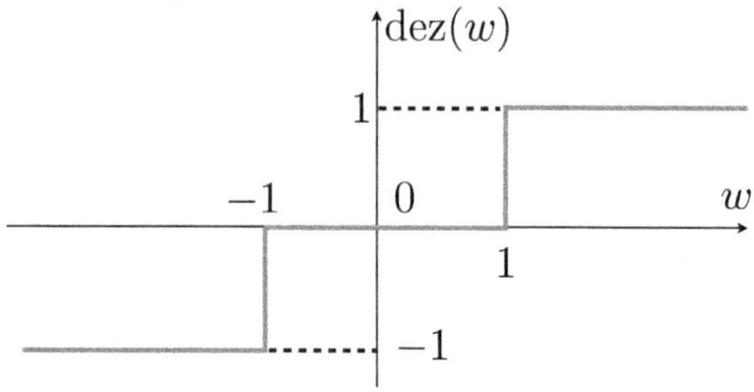

Figure 9.3: Dead-zone function $\mathrm{dez}(w)$

We first consider the case $\eta = 1$ (the normal case). Let u^* denote the L^1 optimal control, and $\boldsymbol{x}^*$ and $\boldsymbol{p}^*$ the associated optimal state and costate, respectively. From the minimum condition in the minimum principle, we have

$$\begin{aligned} u^*(t) &= \operatorname*{arg\,min}_{u\in[-1,1]} H^1(\boldsymbol{x}^*(t), \boldsymbol{p}^*(t), u) \\ &= \operatorname*{arg\,min}_{u\in[-1,1]} \{\boldsymbol{p}^*(t)^\top (A\boldsymbol{x}^*(t) + \boldsymbol{b}u) + |u|\} \\ &= \operatorname*{arg\,min}_{u\in[-1,1]} \{\boldsymbol{p}^*(t)^\top \boldsymbol{b}u + |u|\}. \end{aligned} \tag{9.20}$$

Now, from

$$\boldsymbol{p}^*(t)^\top \boldsymbol{b}u + |u| = \begin{cases} (\boldsymbol{p}^*(t)^\top \boldsymbol{b} + 1)u, & \text{if } \ u \geq 0, \\ (\boldsymbol{p}^*(t)^\top \boldsymbol{b} - 1)u, & \text{if } \ u < 0, \end{cases} \tag{9.21}$$

we have the solution to the minimization problem in (9.20) as

$$u^*(t) = -\mathrm{dez}(\boldsymbol{p}^*(t)^\top \boldsymbol{b}), \tag{9.22}$$

where $\mathrm{dez}(\cdot)$ is the *dead-zone function* defined by

$$\begin{aligned} &\mathrm{dez}(w) \triangleq \begin{cases} -1, & \text{if } \ w < -1, \\ 0, & \text{if } \ -1 < w < 1, \\ 1, & \text{if } \ 1 < w, \end{cases} \\ &\mathrm{dez}(w) \in [-1, 0], \ \text{ if } \ w = -1, \\ &\mathrm{dez}(w) \in [0, 1], \ \text{ if } \ w = 1. \end{aligned} \tag{9.23}$$

Figure 9.3 shows the graph of the dead-zone function.

Exercise 9.1. Show that (9.22) is the solution to the minimization problem (9.20).

If there is a time interval (t_1, t_2) on which $\boldsymbol{p}^*(t)^\top \boldsymbol{b} \equiv \pm 1$ holds, then from (9.23), we cannot uniquely determine $u^*(t)$ on this interval. We call such a time interval a *singular interval.* If an L^1-optimal control problem has a singular interval whose length is positive, then we call the problem a *singular problem.* On the other hand, if

$$\mu(\{t \in [0, T] : |\boldsymbol{p}^*(t)^\top \boldsymbol{b}| = 1\}) = 0 \tag{9.24}$$

holds, then the L^1-optimal control problem is said to be *non-singular.* The following lemma gives a sufficient condition for the non-singularity.

Lemma 9.1. If $(A, \boldsymbol{b})$ in (9.10) is controllable and A is non-singular, then (9.24) holds (i.e., the L^1-optimal control problem is non-singular).

From now on, we say $(A, \boldsymbol{b})$ *is non-singular* if $(A, \boldsymbol{b})$ is controllable and A is non-singular.

Exercise 9.2. Prove Lemma 9.1.

From Lemma (9.1), if $(A, \boldsymbol{b})$ is non-singular, then we have

$$\boldsymbol{p}^*(t)^\top \boldsymbol{b} \neq \pm 1, \quad \text{for almost all} \ \ t \in [0, T]. \tag{9.25}$$

Then, from (9.22) and (9.23), the L^1-optimal control takes values ± 1 or 0 for almost all t in $[0, T]$. We call such a control a *bang-off-bang* control. Figure 9.4 illustrates the bang-off-bang property of L^1-optimal control. We summarize this property as a theorem.

Theorem 9.1. Assume that $(A, \boldsymbol{b})$ is non-singular. Then the L^1-optimal control is bang-off-bang (if it exists).

The bang-off-bang property is important to examine the relation between L^1 and L^0 controls as shown in the next section.

Remark 9.1. The function $\boldsymbol{p}^*(t)^\top \boldsymbol{b}$ is given by

$$\boldsymbol{p}^*(t)^\top \boldsymbol{b} = \left(e^{-A^\top t} \boldsymbol{p}^*(0)\right)^\top \boldsymbol{b} = \boldsymbol{p}^*(0)^\top e^{-At} \boldsymbol{b}, \tag{9.26}$$

from the adjoint equation for $\boldsymbol{p}^*(t)$ in (8.44). Therefore, the function $\boldsymbol{p}^*(t)^\top \boldsymbol{b}$ is continuous and represented by

$$\boldsymbol{p}^*(0)^\top e^{-At} \boldsymbol{b} = \sum_{i=1}^{d} c_i(t) e^{-\lambda_i t}, \tag{9.27}$$

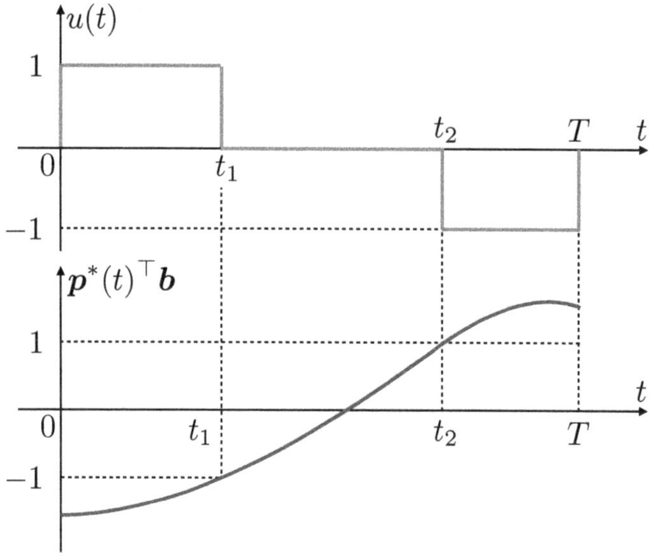

Figure 9.4: L^1-optimal control (bang-off-bang) $u^*(t)$ (top) and function $\boldsymbol{p}^*(t)^\top \boldsymbol{b}$

where λ_i is the i-th eigenvalue of A and $c_i(t)$ is a polynomial with degrees up to d. It follows that the number of switchings in the L^1-optimal control $u^*(t)$ is finite, and the value changes between 1 and 0 or -1 and 0, and never changes between 1 and -1. Therefore, if $u^*(t)$ switches, then there exists a time duration with positive length on which $u^*(t) = 0$, in the non-singular case.

Finally, let us consider the case $\eta = 0$ (the abnormal case). The Hamiltonian is given by

$$H^0(\boldsymbol{x}, \boldsymbol{p}, u) = \boldsymbol{p}^\top (A\boldsymbol{x} + \boldsymbol{b}u). \tag{9.28}$$

Then, the optimal control $u^*(t)$ satisfies

$$\begin{aligned} u^*(t) &= \arg\min_{u\in[-1,1]} H^0(\boldsymbol{x}^*(t), \boldsymbol{p}^*(t), u) \\ &= \arg\min_{u\in[-1,1]} \boldsymbol{p}^*(t)^\top \boldsymbol{b}u \\ &= \begin{cases} -1, & \text{if } \boldsymbol{p}^*(t)^\top \boldsymbol{b} > 0, \\ 1, & \text{if } \boldsymbol{p}^*(t)^\top \boldsymbol{b} < 0, \\ [-1, 1], & \text{if } \boldsymbol{p}^*(t)^\top \boldsymbol{b} = 0. \end{cases} \end{aligned} \tag{9.29}$$

If $(A, \boldsymbol{b})$ is controllable, then $\boldsymbol{p}^*(t)^\top \boldsymbol{b} \not\equiv 0$, and hence the control is *bang-*

bang, taking values of ± 1. With this control, the L^1-optimal value is

$$J_1(u^*) = \int_0^T |u^*(t)|dt = T. \tag{9.30}$$

Moreover, if $T^*(\boldsymbol{\xi}) < T < \infty$, then there exists the minimum-time control $u^*_{\text{time}}(t)$ to achieve $\boldsymbol{x}(T^*(\boldsymbol{\xi})) = \mathbf{0}$. By using this minimum-time control, define the following control:

$$\bar{u}(t) = \begin{cases} u^*_{\text{time}}(t), & \text{if } 0 \leq t \leq T^*(\boldsymbol{\xi}), \\ 0, & \text{if } T^*(\boldsymbol{\xi}) < t \leq T. \end{cases} \tag{9.31}$$

It is easily shown that $\bar{u}$ is a feasible control, that is $\bar{u} \in \mathcal{U}(T, \boldsymbol{\xi})$. Also, with this $\bar{u}$, we have

$$J_1(\bar{u}) = \int_0^T |\bar{u}(t)|dt = \int_0^{T^*(\boldsymbol{\xi})} |u^*_{\text{time}}(t)|dt = T^*(\boldsymbol{\xi}) < T = J_1(u^*). \tag{9.32}$$

Hence, the control $u^*(t)$ can never be L^1 optimal, and hence the case $\eta = 0$ never happens.

The abnormal case ($\eta = 0$) happens when $T = T^*(\boldsymbol{\xi})$. In this case, the set of feasible controls is $\mathcal{U}(T^*(\boldsymbol{\xi}), \boldsymbol{\xi}) = \{u^*_{\text{time}}\}$, a singleton of the minimum-time control, and hence the cost function is meaningless to choose a control from the feasible set. In this book, we do not discuss the abnormal case any further.

9.5 Equivalence Theorem

In this section, we study the equivalence between L^0 and L^1 optimal controls.

The following theorem is a fundamental theorem for the equivalence.

Theorem 9.2. Assume that there exists an L^1-optimal control that is bang-off-bang. Then it is also L^0 optimal.

Proof: Define $J_0(u) \triangleq \|u\|_0$ and $J_1(u) \triangleq \|u\|_1$. From the assumption, there exists an L^1-optimal control u_1^* that is bang-off-bang. Since u_1^* is a feasible control, the set of feasible controls $\mathcal{U}(T, \boldsymbol{\xi})$ is non-empty. Then, for any $u \in \mathcal{U}(T, \boldsymbol{\xi})$ we have

$$J_1(u_1^*) \leq J_1(u) = \int_0^T |u(t)|dt = \int_{\text{supp}(u)} |u(t)|dt \leq \int_{\text{supp}(u)} 1dt = J_0(u). \tag{9.33}$$

Since u_1^* is bang-off-bang, we have

$$J_1(u_1^*) = \int_0^T |u_1^*(t)|dt = \int_{\mathrm{supp}(u_1^*)} 1dt = J_0(u_1^*). \tag{9.34}$$

From (9.33) and (9.34), we have

$$J_0(u_1^*) \leq J_0(u), \quad \forall u \in \mathcal{U}(T, \boldsymbol{\xi}), \tag{9.35}$$

and hence u_1^* minimizes $J_0(u)$. That is, u_1^* is also L^0 optimal. □

From Theorem 9.1, if $(A, \boldsymbol{b})$ is non-singular, then the L^1-optimal control is bang-off-bang, that is, the optimal control $u^*(t)$ takes values 0 or ± 1 for almost all $t \in [0, T]$. From this property, we can obtain the following theorem.

Theorem 9.3. Assume that there exists at least one L^1-optimal control. Assume also that $(A, \boldsymbol{b})$ is non-singular. Then there exists at least one L^0-optimal control, and the set of L^0-optimal controls is identical to the set of L^1-optimal controls.

Proof: Let $\mathcal{U}_0^*$ and $\mathcal{U}_1^*$ be the sets of L^0 and L^1 optimal controls, respectively. From the assumption, $\mathcal{U}_1^*$ is non-empty. Take $u_1^* \in \mathcal{U}_1^*$ arbitrarily. Then, from Theorem 9.1, u_1^* is bang-off-bang. It follows from Theorem 9.2 that $u_1^* \in \mathcal{U}_0^*$, and hence $\mathcal{U}_1^* \subset \mathcal{U}_0^*$.

Then we prove $\mathcal{U}_0^* \subset \mathcal{U}_1^*$. Take $u_0^* \in \mathcal{U}_0^* \subset \mathcal{U}(T, \boldsymbol{\xi})$ arbitrarily. Take also $u_1^* \in \mathcal{U}_1^* \subset \mathcal{U}(T, \boldsymbol{\xi})$ independently. From (9.34) and the L^1 optimality of u_1^*, we have

$$J_0(u_1^*) = J_1(u_1^*) \leq J_1(u_0^*). \tag{9.36}$$

On the other hand, from (9.33) and the L^0 optimality of u_0^*, we have

$$J_1(u_0^*) \leq J_0(u_0^*) \leq J_0(u_1^*). \tag{9.37}$$

From (9.36) and (9.37), we have

$$J_0(u_1^*) = J_1(u_1^*) \leq J_1(u_0^*) \leq J_0(u_0^*) \leq J_0(u_1^*). \tag{9.38}$$

It follows that $J_1(u_1^*) = J_1(u_0^*)$, and u_0^* minimizes $J_1(u)$. That is, we have $u_0^* \in \mathcal{U}_1^*$ and hence $\mathcal{U}_0^* \subset \mathcal{U}_1^*$. □

9.6 Existence of L^0-optimal Control

Here we consider the existence of L^0-optimal control.

9.6.1 L^p-optimal control

To consider the existence of L^0-optimal control, we introduce the L^p-*optimal control* with $p \in (0,1)$. The optimal control problem is described as follows:

Problem 9.3 (L^p-optimal control problem). For the linear time-invariant system

$$\dot{\boldsymbol{x}}(t) = A\boldsymbol{x}(t) + \boldsymbol{b}u(t), \quad t \geq 0, \quad \boldsymbol{x}(0) = \boldsymbol{\xi} \in \mathbb{R}^d, \tag{9.39}$$

find a control $\{u(t) : t \in [0,T]\}$ with $T > 0$ that minimizes

$$J_p(u) = \|u\|_p^p = \int_0^T |u(t)|^p dt \tag{9.40}$$

with $p \in (0,1)$, subject to

$$\boldsymbol{x}(T) = \boldsymbol{0}, \tag{9.41}$$

and

$$\|u\|_\infty \leq 1. \tag{9.42}$$

First, we prove an interesting relation between the L^p norm[2] with $p \in (0,1)$ and the L^0 norm.

Lemma 9.2. Suppose $u \in L^1(0,T)$. Then u is also in $L^p(0,T)$ for any $p \in (0,1)$, and

$$\lim_{p \to 0+} \|u\|_p^p = \|u\|_0. \tag{9.43}$$

Exercise 9.3. Prove Lemma 9.2.

Now, let us look into the L^p-optimal control with $p \in (0,1)$. The Hamiltonian is given by

$$H^\eta(\boldsymbol{x}, \boldsymbol{p}, u) = \boldsymbol{p}^\top (A\boldsymbol{x} + \boldsymbol{b}u) + \eta |u|^p. \tag{9.44}$$

Let us consider the normal case ($\eta = 1$). From the minimum condition in

[2]Strictly speaking, the L^p "norm" with $p \in (0,1)$ is not a proper norm since the triangle inequality does not always hold.

Pontryagin's minimum principle, we have

$$
\begin{aligned}
u^*(t) &= \arg\min_{u\in[-1,1]} H^1(\boldsymbol{x}^*(t), \boldsymbol{p}^*(t), u) \\
&= \arg\min_{u\in[-1,1]} \{\boldsymbol{p}^*(t)^\top \boldsymbol{b} u + |u|^p\} \\
&= \begin{cases} -1, & \text{if } \ \boldsymbol{p}^*(t)^\top \boldsymbol{b} > 1, \\ 0, & \text{if } \ -1 < \boldsymbol{p}^*(t)^\top \boldsymbol{b} < 1, \\ 1, & \text{if } \ \boldsymbol{p}^*(t)^\top \boldsymbol{b} < -1, \\ \{-1, 0\}, & \text{if } \ \boldsymbol{p}^*(t)^\top \boldsymbol{b} = 1, \\ \{0, 1\}, & \text{if } \ \boldsymbol{p}^*(t)^\top \boldsymbol{b} = -1. \end{cases}
\end{aligned}
\tag{9.45}
$$

From this, L^p-optimal control is always *bang-off-bang*. We mention this in the following theorem.

Theorem 9.4. The L^p-optimal control with $p \in (0, 1)$ is bang-off-bang (if it exists).

Also, it is shown that the set of L^p-optimal control is identical to the set of L^0-optimal controls.

Theorem 9.5. Assume that there exists at least one L^p-optimal control with $p \in (0, 1)$. Let $\mathcal{U}_0^*$ and $\mathcal{U}_p^*$ be the sets of L^0 and L^p optimal controls respectively. Then we have

$$
\mathcal{U}_0^* = \mathcal{U}_p^*. \tag{9.46}
$$

Exercise 9.4. Prove Theorem 9.5.

From Theorems 9.4 and 9.5, we have the following theorem.

Theorem 9.6. The L^0-optimal control is bang-off-bang (if it exists).

The difference of Theorem 9.5 from Theorem 9.3 for the L^1-optimal control is that for the L^p optimal control we do not need the assumption of the non-singularity of $(A, \boldsymbol{b})$. This is the key to prove the existence of L^0 optimal control.

9.6.2 Existence theorems

From Theorem 9.5, if we show the existence of L^p-optimal control for some $p \in (0,1)$, then $\mathcal{U}_0^*$ is non-empty, and hence there exists at least one L^0-optimal control. The following theorem is on the existence of L^p-optimal control with $p > 0$.[3]

Theorem 9.7. Suppose that the initial state $\boldsymbol{\xi} \in \mathbb{R}^d$ and the time $T > 0$ are chosen such that $\boldsymbol{\xi} \in \mathcal{R}(T)$. Then, there exists an L^p-optimal control with $p > 0$.

Proof: Assume $\boldsymbol{\xi} \in \mathbb{R}^d$ and $T > 0$ satisfy $\boldsymbol{\xi} \in \mathcal{R}(T)$. Then there exists a feasible control $u \in \mathcal{U}(T, \boldsymbol{\xi})$, and hence the feasible set $\mathcal{U}(T, \boldsymbol{\xi})$ is non-empty. Define

$$J_p^* \triangleq \inf\{\|u\|_p^p : u \in \mathcal{U}(T, \boldsymbol{\xi})\}. \tag{9.47}$$

Since $u \in \mathcal{U}(T, \boldsymbol{\xi})$ satisfies $\|u\|_\infty \leq 1$, we have $J_p^* < \infty$. Then, from the definition of J_p^*, there exists a sequence $\{u_l\}_{l \in \mathbb{N}} \subset \mathcal{U}(T, \boldsymbol{\xi})$ such that

$$\lim_{l \to \infty} \|u_l\|_p^p = J_p^*. \tag{9.48}$$

Now, since $u_l \in \mathcal{U}(T, \boldsymbol{\xi})$, we have

$$\|u_l\|_\infty \leq 1, \tag{9.49}$$

and hence $\{u_l\}_{l \in \mathbb{N}} \subset \mathcal{B}_\infty \triangleq \{u \in L^\infty(0,T) : \|u\|_\infty \leq 1\}$. It is known that the unit ball $\mathcal{B}_\infty$ is sequentially compact in the weak* topology of $L^\infty(0,T)$ [92, Theorem A.9]. That is, there exists a subsequence $\{u_{l'}\}_{l' \in S}$, $S \subset \mathbb{N}$, such that there exists $u_\infty \in \mathcal{B}_\infty$ and

$$\lim_{l' \to \infty} \int_0^T f(t)(u_{l'}(t) - u_\infty(t))dt = 0, \tag{9.50}$$

for any $f \in L^1(0,T)$. Now, since $u_{l'} \in \mathcal{U}(T, \boldsymbol{\xi})$, we have

$$\boldsymbol{\xi} = -\int_0^T e^{-At}\boldsymbol{b}u_{l'}(t)dt, \quad \forall l' \in S. \tag{9.51}$$

On the other hand, from (9.50) with $f(t) = e^{-At}\boldsymbol{b}$, we have

$$\lim_{l' \to \infty} \int_0^T e^{-At}\boldsymbol{b}u_{l'}(t)dt = \int_0^T e^{-At}\boldsymbol{b}u_\infty(t)dt. \tag{9.52}$$

[3] To show the existence of L^0-optimal control, we just need to prove the existence of L^p-optimal control with $p \in (0,1)$. However, Theorem 9.7 gives more general result.

That is,

$$\boldsymbol{\xi} = -\int_0^T e^{-At}\boldsymbol{b}u_\infty(t)dt. \tag{9.53}$$

Also since $u_\infty \in \mathcal{B}_\infty$, we have $\|u_\infty\|_\infty \leq 1$. Therefore, $u_\infty \in \mathcal{U}(T, \boldsymbol{\xi})$.

Next, from (9.50) with $f(t) = \text{sgn}(u_{l'}(t) - u_\infty(t))$, [4] we have

$$\lim_{l'\to\infty} \int_0^T |u_{l'}(t) - u_\infty(t)|dt = 0. \tag{9.54}$$

Then, the following inequalities hold:

$$\begin{aligned} \left|\|u_{l'}\|_p^p - \|u_\infty\|_p^p\right| &\leq \int_0^T \left||u_{l'}(t)|^p - |u_\infty(t)|^p\right|dt \\ &\leq \int_0^T \left|u_{l'}(t)^p - u_\infty(t)^p\right|^p dt \\ &\leq \left(\int_0^T |u_{l'}(t) - u_\infty(t)|dt\right)^p T^{1-p}, \end{aligned} \tag{9.55}$$

where the third inequality is from Hölder's inequality. It follows from (9.54) and (9.55) that

$$\lim_{l'\to\infty} \|u_{l'}\|_p^p = \|u_\infty\|_p^p. \tag{9.56}$$

The left-hand side of the above equation is equivalent to J_p^* by definition, and hence $\|u_\infty\|_p^p = J_p^*$. That is, $u_\infty \in \mathcal{U}(T, \xi)$ is an L^p-optimal control. □

From Theorem 9.7 and Theorem 9.5, we have the following theorem.

Theorem 9.8 (Existence of L^0 optimal control)**.** If $\boldsymbol{\xi} \in \mathcal{R}(T)$, then there exists an L^0-optimal control.

The condition of $\boldsymbol{\xi} \in \mathcal{R}(T)$ is equivalent to $\boldsymbol{\xi} \in \mathcal{R}$ and $T \geq T^*(\boldsymbol{\xi})$. Hence, we have the following lemma.

Lemma 9.3. Let u^* be the L^0-optimal control with $\boldsymbol{\xi} \in \mathcal{R}(T)$. Then $\|u^*\|_0 \leq T^*(\boldsymbol{\xi})$.

Proof: This is easily shown by considering a feasible control in (9.31). □

[4] Here sgn(·) is the sign function defined by

$$\text{sgn}(v) \triangleq \begin{cases} -1, & \text{if } v < 0, \\ 0, & \text{if } v = 0, \\ 1, & \text{if } v > 0. \end{cases}$$

9.7 Rocket Control Example

Here we compute the maximum hands-off control of the rocket considered in Example 8.1 (p. 176) in the previous chapter. We assume the mass $m = 1$ for simplicity.

We now compute the L^1-optimal control. From (9.19), the Hamiltonian with $\eta = 1$ is given by

$$H^1(\boldsymbol{x}, \boldsymbol{p}, u) = \boldsymbol{p}^\top \left(\begin{bmatrix} 0 & 1 \\ 0 & 0 \end{bmatrix} \boldsymbol{x} + \begin{bmatrix} 0 \\ 1 \end{bmatrix} u \right) + |u| = p_1 x_1 + p_2 u + |u|, \quad (9.57)$$

where $\boldsymbol{p} = (p_1, p_2)$. Let u^* denote the L^1-optimal control, and $\boldsymbol{x}^* = (x_1^*, x_2^*)$, $\boldsymbol{p}^* = (p_1^*, p_2^*)$ the associated optimal state and costate, respectively. From (9.22), the L^1-optimal control $u^*(t)$ satisfies

$$u^*(t) = -\mathrm{dez}(p_2^*(t)), \quad (9.58)$$

where $\mathrm{dez}(\cdot)$ is the dead-zone function defined in (9.23) (see also Figure 9.3).

Then, the adjoint equation of the costate $\boldsymbol{p}^*(t)$ becomes

$$\begin{bmatrix} \dot{p}_1^*(t) \\ \dot{p}_2^*(t) \end{bmatrix} = - \begin{bmatrix} 0 & 1 \\ 0 & 0 \end{bmatrix}^\top \begin{bmatrix} p_1^*(t) \\ p_2^*(t) \end{bmatrix} = \begin{bmatrix} 0 \\ -p_1^*(t) \end{bmatrix}. \quad (9.59)$$

The solution of this differential equation is given by

$$p_1^*(t) = \pi_1, \quad p_2^*(t) = \pi_2 - \pi_1 t, \quad (9.60)$$

where

$$\pi_1 = p_1^*(0), \quad \pi_2 = p_2^*(0). \quad (9.61)$$

It follows from (9.60) that if $\pi_1 \neq 0$ then $p_2^*(t)$ is a first-order linear function of t and hence $p_2^*(t)$ is monotone. Therefore, from (9.58) and the definition of the dead-zone function (9.23), switching occurs at most twice, and the value changes between -1 and 0, or between 0 and 1. From this observation, the L^1-optimal control is given as follows (for details, refer to [3, Section 8.5]).

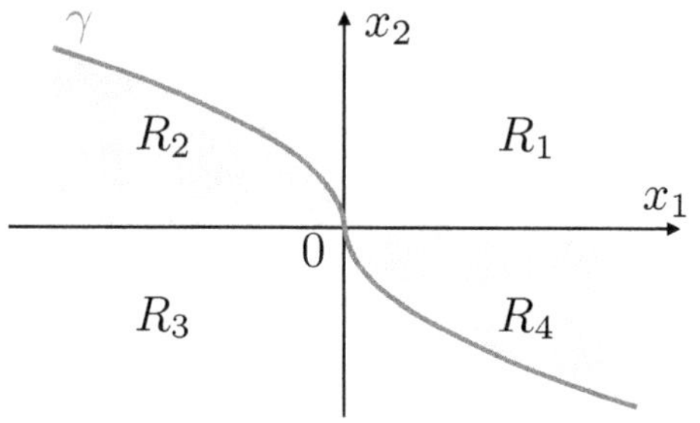

Figure 9.5: Curve γ (thick solid line) and regions R_1, R_2, R_3, and R_4

Define the following regions (see Figure 9.5):

$$\begin{aligned}\gamma &= \left\{(x_1, x_2) \in \mathbb{R}^2 : x_1 = -x_2|x_2|/2\right\},\\ R_1 &= \left\{(x_1, x_2) \in \mathbb{R}^2 : x_1 > -x_2^2/2,\ x_2 \geq 0\right\},\\ R_2 &= \left\{(x_1, x_2) \in \mathbb{R}^2 : x_1 < -x_2^2/2,\ x_2 > 0\right\},\\ R_3 &= \left\{(x_1, x_2) \in \mathbb{R}^2 : x_1 < x_2^2/2,\ x_2 \leq 0\right\},\\ R_4 &= \left\{(x_1, x_2) \in \mathbb{R}^2 : x_1 > x_2^2/2,\ x_2 < 0\right\}.\end{aligned} \tag{9.62}$$

Also, define the following two regions:

$$\begin{aligned}V_- &= \left\{(x_1, x_2) \in \mathbb{R}^2 : -x_2/2 - x_1/x_2 \geq T\right\},\\ V_+ &= \left\{(x_1, x_2) \in \mathbb{R}^2 : x_2/2 - x_1/x_2 \geq T\right\}.\end{aligned} \tag{9.63}$$

Then, the L^1-optimal control is given as follows.

1. If $(\xi_1, \xi_2) \in R_1$ or $(\xi_1, \xi_2) \in R_4 \cap V_-$, then the optimal control is given by

$$u^*(t) = \begin{cases} -1, & \text{if } 0 \leq t < t_1,\\ 0, & \text{if } t_1 \leq t < t_2,\\ 1, & \text{if } t_2 \leq t \leq T,\end{cases} \tag{9.64}$$

where

$$\begin{aligned} t_1 &= \frac{T + \xi_2 - \sqrt{(T - \xi_2)^2 - 4\xi_1 - 2\xi_2^2}}{2},\\ t_2 &= \frac{T + \xi_2 + \sqrt{(T - \xi_2)^2 - 4\xi_1 - 2\xi_2^2}}{2}.\end{aligned} \tag{9.65}$$

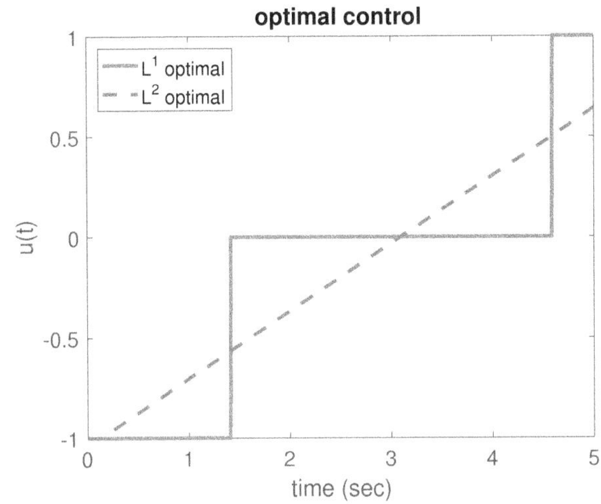

Figure 9.6: L^1-optimal control (solid line) and L^2-optimal control (dashed line)

2. If $(\xi_1, \xi_2) \in R_3$ or $(\xi_1, \xi_2) \in R_2 \cap V_+$ then the optimal control is given by

$$u^*(t) = \begin{cases} 1, & \text{if } 0 \le t < t_3, \\ 0, & \text{if } t_3 \le t < t_4, \\ -1, & \text{if } t_4 \le t \le T, \end{cases} \tag{9.66}$$

where

$$\begin{aligned} t_3 &= \frac{T - \xi_2 - \sqrt{(T+\xi_2)^2 + 4\xi_1 - 2\xi_2^2}}{2}, \\ t_4 &= \frac{T - \xi_2 + \sqrt{(T+\xi_2)^2 + 4\xi_1 - 2\xi_2^2}}{2}. \end{aligned} \tag{9.67}$$

3. If $(\xi_1, \xi_2) \in \gamma$ then the optimal control is given by

$$u^*(t) = \begin{cases} -\text{sgn}(\xi_2), & \text{if } 0 \le t < |\xi_2|, \\ 0, & \text{if } |\xi_2| \le t \le T. \end{cases} \tag{9.68}$$

4. If $(\xi_1, \xi_2) \in R_4 \cap (V_-)^c$ or $(\xi_1, \xi_2) \in R_2 \cap (V_+)^c$, [5] then the control problem is *singular*, and the optimal control cannot be uniquely determined from the minimum principle.

[5] $(\cdot)^c$ denotes the complement set.

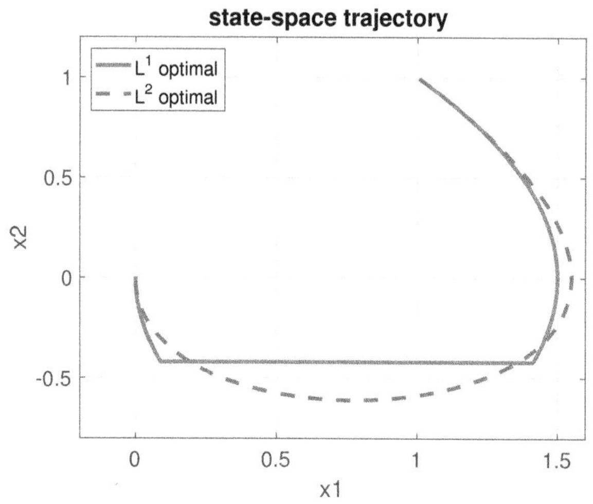

Figure 9.7: Optimal state trajectory $(x_1^*(t), x_2^*(t))$: L^1-optimal control (solid line) and L^2-optimal control (dashed line)

Figure 9.6 shows the L^1-optimal control with the final time $T = 5$ and the initial state $(\xi_1, \xi_2) = (1, 1) \in R_1$. Figure 9.7 shows the associated optimal state trajectory $\{(x_1^*(t), x_2^*(t)) : 0 \leq t \leq 5\}$. In these figures, we also show the results of L^2-optimal control that minimizes the L^2 cost function

$$\|u\|_2^2 = \int_0^T |u(t)|^2 dt, \tag{9.69}$$

among the feasible controls. From Figure 9.6, we can see that the L^1-optimal control is sparse while the L^2-optimal control is not. In fact, the L^1-optimal control is bang-off-bang, and hence this is equivalent to L^0-optimal control. That is, the L^1-optimal control has the maximum length of time duration on which the control is exactly zero. From (9.65), this time length is given by

$$[t_1, t_2] = [3 - \sqrt{10}/2, 3 + \sqrt{10}/2] \approx [1.4189, 4.5811], \tag{9.70}$$

and the L^0 norm of the L^1-optimal control u^* is $\|u^*\|_0 = \sqrt{10} \approx 3.1623$. On this time duration, the state trajectory $(x_1^*(t), x_2^*(t))$ is parallel to the x_1 axes. Since x_1 is the portion and x_2 is the velocity of the rocket, this state trajectory means that the rocket moves at a constant velocity. The rocket consumes no fuel on this time duration, and hence we can cut fuel consumptions and also we can reduce CO2 emissions etc. That is, the control is *green*. It is clear that L^2-optimal control does not have such a nice property of sparsity.

9.8 Further Readings

For the L^1-optimal control (minimum-fuel control), the most detailed information can be obtained from the classical book by Athans and Falb [3]. The equivalence theorem between L^0 and L^1 optimal controls was first proved in [104], [105]. For the equivalence, we need the assumption of non-singularity of A. In the case of singular A, we can adopt non-convex surrogate functions, such as the L^p norm with $p \in (0, 1)$ and the minimax concave penalty, with which we can show the equivalence to the L^0-optimal control [55], [59], [63].

In this book, we consider linear systems, but the equivalence property holds for non-linear systems of the following type:

$$\dot{\boldsymbol{x}}(t) = f(\boldsymbol{x}(t)) + g(\boldsymbol{x}(t))u(t), \quad t \geq 0. \tag{9.71}$$

See [104], [105] for details.

Necessary conditions of the L^0-optimal control are also obtained in [23] by the non-smooth version of Pontryagin's minimum (or maximum) principle [28]. For the theory of L^p spaces, see [79], [133], [152].

For feedback control implementation of maximum hands-off control, one can adopt the model predictive control [32], [64], [113] and the self-triggered control [105].

Chapter 10

Numerical Optimization by Time Discretization

As we have seen in Section 9.7, the L^1-optimal control is obtained in a closed form when the plant is very simple, as the double integrator. However, for general systems described by

$$\dot{\boldsymbol{x}}(t) = A\boldsymbol{x}(t) + \boldsymbol{b}u(t), \quad t \geq 0, \quad \boldsymbol{x}(0) = \boldsymbol{\xi} \in \mathbb{R}^d, \tag{10.1}$$

we need to rely on numerical computation to obtain the optimal control. In this chapter, we introduce the method of time discretization to numerically obtain the L^1-optimal control.

Key ideas of Chapter 10

- By time discretization, the L^1-optimal control problem (Problem 9.2) is reduced to a finite-dimensional ℓ^1 optimization problem.
- In time discretization, the control is assumed to be piecewise constant by a zero-order hold.
- The reduced ℓ^1 optimization can be efficiently solved by ADMM.

10.1 Time Discretization

First, we discretize the time interval $[0, T]$ into n subintervals as

$$[0, T] = [0, h) \cup [h, 2h) \cup \cdots \cup [nh - h, nh], \tag{10.2}$$

where $h > 0$ is the sampling time and $n \in \mathbb{N}$ is the number of subintervals such that $T = nh$.

On each subinterval, we assume the control $u(t)$ is constant. More precisely, we assume the control is given by

$$u(t) = u(kh) = u_{\mathrm{d}}[k], \quad t \in [kh, (k+1)h), \quad k = 0, 1, 2, \ldots, n-1. \tag{10.3}$$

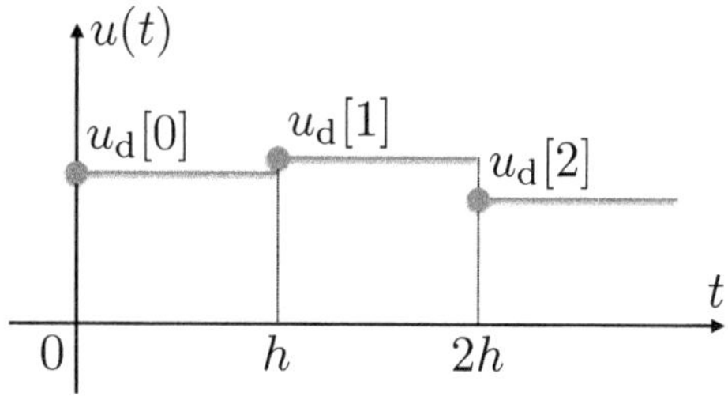

Figure 10.1: Zero-order hold output of discrete-time signal $\{u_{\rm d}[k]\}$

This is the output of a *zero-order hold* of a discrete-time signal

$$u_{\rm d} \triangleq \{u_{\rm d}[0], u_{\rm d}[1], \ldots, u_{\rm d}[n-1]\}. \tag{10.4}$$

This assumption is actually reasonable for networked digital control systems where control values are computed in a digital computer, transmitted through a wireless communication network, and applied to an actuator through a D/A converter. The zero-order hold is the simplest model of a D/A converter.

Let us compute the state transition under the zero-order assumption on the control. The solution to the state-space equation in (10.1) is given by (see Exercise 8.1 on p. 176)

$$\boldsymbol{x}(t_1) = e^{A(t_1-t_0)}\boldsymbol{x}(t_0) + \int_{t_0}^{t_1} e^{A(t_1-\tau)}\boldsymbol{b}u(\tau)d\tau, \tag{10.5}$$

where $0 \leq t_0 \leq t_1$. Take

$$t_0 = kh, \quad t_1 = kh + h, \quad k \in \{0, 1, 2, \ldots, n-1\}. \tag{10.6}$$

Then from (10.5) we have

$$\begin{aligned} \boldsymbol{x}(kh+h) &= e^{Ah}\boldsymbol{x}(kh) + \int_{kh}^{kh+h} e^{A(kh+h-\tau)}\boldsymbol{b}u(\tau)d\tau \\ &= e^{Ah}\boldsymbol{x}(kh) + \int_0^h e^{A(h-t)}\boldsymbol{b}u(t+kh)dt. \end{aligned} \tag{10.7}$$

Define

$$\boldsymbol{x}_{\rm d}[k] \triangleq \boldsymbol{x}(kh), \quad u_{\rm d}[k] \triangleq u(kh), \quad k = 0, 1, \ldots, n-1, \tag{10.8}$$

and

$$\boldsymbol{x}_{\rm d}[n] \triangleq \boldsymbol{x}(T). \tag{10.9}$$

From the zero-order-hold assumption (10.3), the control $u(t)$ takes a constant value $u_{\mathrm{d}}[k] = u(kh)$ on the subinterval $[kh, kh+h)$ as shown in Figure 10.1. Then from (10.7) we have

$$\boldsymbol{x}_{\mathrm{d}}[k+1] = e^{Ah}\boldsymbol{x}_{\mathrm{d}}[k] + \left(\int_0^h e^{A(h-t)}\boldsymbol{b}\,dt\right) u_{\mathrm{d}}[k]. \tag{10.10}$$

It follows that the differential equation (10.1) is transformed into the following difference equation:

$$\boldsymbol{x}_{\mathrm{d}}[k+1] = A_{\mathrm{d}}\boldsymbol{x}_{\mathrm{d}}[k] + \boldsymbol{b}_{\mathrm{d}}u_{\mathrm{d}}[k], \quad k = 0, 1, \ldots, n-1, \tag{10.11}$$

where

$$A_{\mathrm{d}} \triangleq e^{Ah}, \quad \boldsymbol{b}_{\mathrm{d}} \triangleq \int_0^h e^{At}\boldsymbol{b}\,dt. \tag{10.12}$$

Next, define the control vector

$$\boldsymbol{u} \triangleq \begin{bmatrix} u_{\mathrm{d}}[0] \\ u_{\mathrm{d}}[1] \\ \vdots \\ u_{\mathrm{d}}[n-1] \end{bmatrix} \in \mathbb{R}^n. \tag{10.13}$$

By using this, the terminal state $\boldsymbol{x}(T)$ is described as

$$\boldsymbol{x}(T) = \boldsymbol{x}_{\mathrm{d}}[n] = -\boldsymbol{\zeta} + \Phi\boldsymbol{u}, \tag{10.14}$$

where

$$\Phi \triangleq \begin{bmatrix} A_{\mathrm{d}}^{n-1}\boldsymbol{b}_{\mathrm{d}} & A_{\mathrm{d}}^{n-2}\boldsymbol{b}_{\mathrm{d}} & \ldots & \boldsymbol{b}_{\mathrm{d}} \end{bmatrix}, \quad \boldsymbol{\zeta} \triangleq -A_{\mathrm{d}}^{n}\boldsymbol{\xi}. \tag{10.15}$$

Exercise 10.1. Show the equation (10.14) by solving the difference equation (10.11).

10.2 Controllability of Discretized Systems

The discrete-time system (10.11) is called the *zero-order-hold discretization* or *step-invariant discretization* of the continuous-time system (10.1). Figure 10.2 shows the zero-order hold discretization of (10.1). In this figure $\mathcal{H}_h$ is the *zero-order hold* with sampling time h, which outputs a constant value $u_{\mathrm{d}}[k] = u(kh)$ over $[kh, (k+1)h)$, $k = 0, 1, 2, \ldots$ (see Figure 10.1). Also, $\mathcal{S}_h$ is the *ideal sampler* that outputs the sampled data $\boldsymbol{x}_{\mathrm{d}}[k] = \boldsymbol{x}(kh)$, $k = 0, 1, 2, \ldots$ of the continuous-time signal $\boldsymbol{x}(t)$.

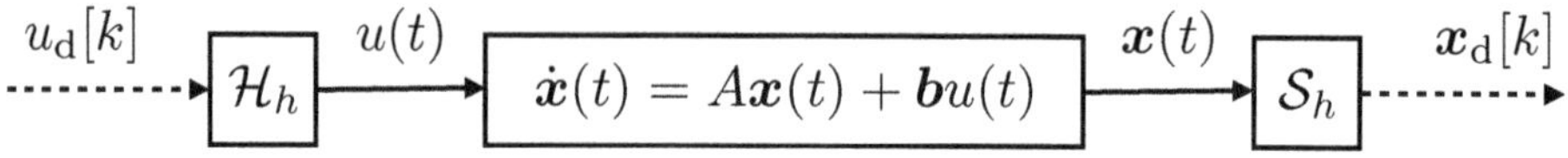

Figure 10.2: Zero-order-hold discretization: continuous-time system $\dot{x}(t) = Ax(t) + bu(t)$ is discretized by zero-order hold $\mathcal{H}_h$ and ideal sampler $\mathcal{S}_h$ with sampling time h.

As discussed above, the discrete-time system from $u_\mathrm{d}[k]$ to $\boldsymbol{x}_\mathrm{d}[k]$ in Figure 10.2 is a linear time-invariant discrete-time system as in (10.11). Then, under this discretization, the stability is preserved; if A is stable, that is, if the eigenvalues of A have non-positive real parts then A_d is Schur stable, that is, the eigenvalues of A_d lie in the closed unit circle in $\mathbb{C}$. This is easily shown from the spectral mapping theorem: the set of the eigenvalues of $A_\mathrm{d} = e^{Ah}$ is given by $\{e^{\lambda_1 h}, \ldots, e^{\lambda_d h}\}$, where λ_i is the i-th eigenvalue of A.

On the other hand, we cannot say the controllability is not always preserved under the zero-order-hold discretization. To discuss this, we introduce the concept of *pathological sampling.*

Definition 10.1 (pathological sampling). Let $\lambda(A)$ be the set of eigenvalues of A. The sampling time $h > 0$ is said to be *pathological* if there exist $\lambda_1, \lambda_2 \in \lambda(A)$ such that

1. $\lambda_1 \neq \lambda_2$,
2. $\mathrm{Re}\,\lambda_1 = \mathrm{Re}\,\lambda_2$,
3. there exists $k \in \{\pm 1, \pm 2, \ldots\}$ such that

$$\mathrm{Im}\,\lambda_1 - \mathrm{Im}\,\lambda_2 = \frac{2\pi k}{h}. \tag{10.16}$$

Intuitively, pathological sampling synchronizes an oscillation mode in the plant. The following illustrates pathological sampling.

Example 10.1. Let us consider a linear system

$$\ddot{y}(t) = -y(t), \quad y(0) = 0, \quad \dot{y}(0) = 1. \tag{10.17}$$

Then the solution of this differential equation is given by

$$y(t) = \sin t. \tag{10.18}$$

If we sample this output with sampling period $h = \pi$, then we have

$$y(kh) = \sin kh = 0, \quad k = 0, 1, 2, \dots \tag{10.19}$$

This is an example of pathological sampling. The state-space representation of (10.17) is given by

$$\frac{d}{dt}\begin{bmatrix} x_1(t) \\ x_2(t) \end{bmatrix} = \begin{bmatrix} 0 & 1 \\ -1 & 0 \end{bmatrix}\begin{bmatrix} x_1(t) \\ x_2(t) \end{bmatrix}, \quad \begin{bmatrix} x_1(0) \\ x_2(0) \end{bmatrix} = \begin{bmatrix} 0 \\ 1 \end{bmatrix}, \tag{10.20}$$

where $x_1(t) \triangleq y(t)$ and $x_2(t) \triangleq \dot{y}(t)$. Then the matrix

$$A = \begin{bmatrix} 0 & 1 \\ -1 & 0 \end{bmatrix} \tag{10.21}$$

has two eigenvalues $\lambda_{\pm} = \pm \mathrm{j}$ satisfying

$$\operatorname{Im} \lambda_{+} - \operatorname{Im} \lambda_{-} = 2 = \frac{2\pi k}{h}, \tag{10.22}$$

with $k = 1$. Therefore, $h = \pi$ is certainly pathological. □

When the sampling is non-pathological, then the controllability is preserved as shown in the following theorem.

Theorem 10.1. Assume that the sampling time h is non-pathological. Then, $(A, \boldsymbol{b})$ is controllable if and only if $(A_{\mathrm{d}}, \boldsymbol{b}_{\mathrm{d}})$ is controllable.

The proof is found in [26].

10.3 Reduction to Finite-dimensional Optimization

Now we reduce the L^1-optimal control problem (Problem 9.2, p. 201) into a finite-dimensional ℓ^1 optimization problem by the time discretization.

First, the constraint on the magnitude of control $\|u\|_{\infty} \leq 1$ is equivalently written by

$$|u_{\mathrm{d}}[k]| \leq 1, \quad \forall k \in \{0, 1, 2, \dots, n-1\}, \tag{10.23}$$

under the zero-order-hold assumption (10.3). Let us denote by $\|\boldsymbol{u}\|_{\ell^\infty}$ the ℓ^∞ norm of a vector $\boldsymbol{u}$ (see (2.29) in Chapter 2). Then the above inequality is equivalent to

$$\|\boldsymbol{u}\|_{\ell^\infty} \leq 1. \tag{10.24}$$

Next, under the zero-order-hold assumption, the L^1 cost function becomes

$$\begin{aligned} J_1(u) &= \int_0^T |u(t)|dt \\ &= \sum_{k=0}^{n-1} \int_{kh}^{(k+1)h} |u(t)|dt \\ &= \sum_{k=0}^{n-1} \int_{kh}^{(k+1)h} |u_{\mathrm{d}}[k]|dt \\ &= \sum_{k=0}^{n-1} |u_{\mathrm{d}}[k]|h \\ &= h\|\boldsymbol{u}\|_{\ell^1}. \end{aligned} \tag{10.25}$$

Now the L^1-optimal control problem (Problem 9.2) is reduced to the following finite-dimensional ℓ^1 optimization problem:

$$\underset{\boldsymbol{u}\in\mathbb{R}^n}{\text{minimize}} \ \|\boldsymbol{u}\|_{\ell^1} \ \text{ subject to } \ \Phi\boldsymbol{u} = \boldsymbol{\zeta}, \ \|\boldsymbol{u}\|_{\ell^\infty} \le 1. \tag{10.26}$$

This optimization problem is a *convex* optimization since the cost function (the ℓ^1 norm) is a convex function, and the constraint set

$$\mathcal{C} \triangleq \{\boldsymbol{u} \in \mathbb{R}^n : \Phi\boldsymbol{u} = \boldsymbol{\zeta}, \ \|\boldsymbol{u}\|_{\ell^\infty} \le 1\} \tag{10.27}$$

is a convex set in $\mathbb{R}^n$. We can easily solve this problem by using CVXPY with Python (see Section 3.3 in Chapter 3, p. 54). A Python program to solve the ℓ^1 optimization (10.26) using CVXPY is given in Section 10.6.1.

10.4 Fast Algorithm by ADMM

If the order d of the system (10.1) and the number n for discretization are not so large, you can obtain a solution easily by CVXPY. However, in real systems, the numerical optimization algorithm should be implemented in a microcomputer, which often has just a cheap computational ability and is hard to run CVXPY. Also, if you want to use the control in a feedback loop, then you must solve the problem in *real time*. In such a case, we need to implement a fast and simple algorithm for the specific ℓ^1 optimization problem (10.26). For this purpose, we can use the efficient algorithms studied in Chapter 4. In particular, we here use ADMM (Alternating Direction Method of Multipliers) studied in Section 4.5 to solve (10.26).

First, define the unit ball $\mathcal{C}_1 \subset \mathbb{R}^n$ with the ℓ^∞ norm by

$$\mathcal{C}_1 \triangleq \{\boldsymbol{u} \in \mathbb{R}^n : \|\boldsymbol{u}\|_{\ell^\infty} \leq 1\}. \tag{10.28}$$

Also, let $\mathcal{C}_2$ be a singleton of $\boldsymbol{\zeta} \in \mathbb{R}^d$, that is,

$$\mathcal{C}_2 \triangleq \{\boldsymbol{\zeta}\}. \tag{10.29}$$

Define the indicator functions of the sets $\mathcal{C}_1$ and $\mathcal{C}_2$ respectively by

$$I_{\mathcal{C}_1}(\boldsymbol{u}) \triangleq \begin{cases} 0, & \text{if } \|\boldsymbol{u}\|_{\ell^\infty} \leq 1, \\ \infty, & \text{if } \|\boldsymbol{u}\|_{\ell^\infty} > 1, \end{cases} \tag{10.30}$$

$$I_{\mathcal{C}_2}(\boldsymbol{x}) \triangleq \begin{cases} 0, & \text{if } \boldsymbol{x} = \boldsymbol{\zeta}, \\ \infty, & \text{if } \boldsymbol{x} \neq \boldsymbol{\zeta}. \end{cases} \tag{10.31}$$

Then the optimization problem (10.26) is equivalently described by

$$\underset{\boldsymbol{u}\in\mathbb{R}^n}{\text{minimize}} \ \{\|\boldsymbol{u}\|_{\ell^1} + I_{\mathcal{C}_1}(\boldsymbol{u}) + I_{\mathcal{C}_2}(\Phi\boldsymbol{u})\}. \tag{10.32}$$

Next, define new variables $\boldsymbol{z}_0, \boldsymbol{z}_1 \in \mathbb{R}^n, \boldsymbol{z}_2 \in \mathbb{R}^d$ by

$$\boldsymbol{z}_0 = \boldsymbol{z}_1 = \boldsymbol{u}, \quad \boldsymbol{z}_2 = \Phi\boldsymbol{u}. \tag{10.33}$$

Then the problem (10.32) becomes

$$\underset{\boldsymbol{u}\in\mathbb{R}^n, \boldsymbol{z}\in\mathbb{R}^\nu}{\text{minimize}} \ \{\|\boldsymbol{z}_0\|_{\ell^1} + I_{\mathcal{C}_1}(\boldsymbol{z}_1) + I_{\mathcal{C}_2}(\boldsymbol{z}_2)\} \ \text{ subject to } \ \boldsymbol{z} = \Psi\boldsymbol{u}, \tag{10.34}$$

where $\nu \triangleq 2n + d$, and

$$\boldsymbol{z} \triangleq \begin{bmatrix} \boldsymbol{z}_0 \\ \boldsymbol{z}_1 \\ \boldsymbol{z}_2 \end{bmatrix} \in \mathbb{R}^\nu, \quad \Psi \triangleq \begin{bmatrix} I \\ I \\ \Phi \end{bmatrix} \in \mathbb{R}^{\nu \times n}. \tag{10.35}$$

Defining two functions f_1 and f_2 by

$$f_1(\boldsymbol{u}) \triangleq 0, \quad f_2(\boldsymbol{z}) \triangleq \|\boldsymbol{z}_0\|_{\ell^1} + I_{\mathcal{C}_1}(\boldsymbol{z}_1) + I_{\mathcal{C}_2}(\boldsymbol{z}_2) \tag{10.36}$$

we finally obtain the standard optimization problem for ADMM (see (4.99), p. 90):

$$\underset{\boldsymbol{u}\in\mathbb{R}^n, \boldsymbol{z}\in\mathbb{R}^\nu}{\text{minimize}} \ f_1(\boldsymbol{u}) + f_2(\boldsymbol{z}) \ \text{ subject to } \ \boldsymbol{z} = \Psi\boldsymbol{u}, \tag{10.37}$$

for which the ADMM algorithm is given by (see Section 4.5.1, p. 90)

$$\boldsymbol{u}[k+1] := \underset{\boldsymbol{u}\in\mathbb{R}^n}{\arg\min} \left\{ f_1(\boldsymbol{u}) + \frac{1}{2\gamma}\|\Psi\boldsymbol{u} - \boldsymbol{z}[k] + \boldsymbol{v}[k]\|_{\ell^2}^2 \right\}, \tag{10.38}$$

$$\boldsymbol{z}[k+1] := \text{prox}_{\gamma f_2}(\Psi\boldsymbol{u}[k+1] + \boldsymbol{v}[k]), \tag{10.39}$$

$$\boldsymbol{v}[k+1] := \boldsymbol{v}[k] + \Psi\boldsymbol{u}[k+1] - \boldsymbol{z}[k+1]. \tag{10.40}$$

Let us compute the functions in (10.38)–(10.40). First, since $f_1 = 0$, the first step (10.38) is minimization of a quadratic function, and it is reduced to the following linear transformation:

$$\begin{aligned} \boldsymbol{u}[k+1] &= \arg\min_{\boldsymbol{u}\in\mathbb{R}^n} \left\{ \frac{1}{2\gamma} \|\Psi\boldsymbol{u} - \boldsymbol{z}[k] + \boldsymbol{v}[k]\|_{\ell^2}^2 \right\} \\ &= (\Psi^\top\Psi)^{-1}\Psi^\top(\boldsymbol{z}[k] - \boldsymbol{v}[k]). \end{aligned} \tag{10.41}$$

Note that $\Psi^\top\Psi = 2I + \Phi^\top\Phi$ is non-singular and the matrix

$$M \triangleq (\Psi^\top\Psi)^{-1}\Psi^\top \tag{10.42}$$

can be computed off-line (i.e., outside the iteration).

The size of $\Psi^\top\Psi$ is $n \times n$, and if the number n of time discretization is very large, then the computation of the inversion may take large computational time. In this case, we can adopt the *matrix inversion lemma*

$$(X + UYV)^{-1} = X^{-1} - X^{-1}U(Y^{-1} + VX^{-1}U)^{-1}VX^{-1}. \tag{10.43}$$

By this, the inverse matrix $(\Psi^\top\Psi)^{-1}$ can be rewritten as

$$(\Psi^\top\Psi)^{-1} = (2I + \Phi^\top\Phi)^{-1} = \frac{1}{2}I - \frac{1}{2}\Phi^\top(2I + \Phi\Phi^\top)^{-1}\Phi. \tag{10.44}$$

This requires inversion of matrix $2I + \Phi\Phi^\top$ of size $d \times d$, and if $d \ll n$ then the computational time can be significantly reduced.

The second step (10.39) in the ADMM algorithm can be split into three simple optimization problems with variables $\boldsymbol{z}_0$, $\boldsymbol{z}_1$, and $\boldsymbol{z}_2$ defined in (10.35). For the variable $\boldsymbol{z}_0$, we use the proximal operator of the ℓ^1 norm, which is the *soft-thresholding operator* defined in (4.46) (see also Figure 4.8 on p .74). That is, the i-th element of $\mathrm{prox}_{\gamma\|\cdot\|_{\ell^1}}(\boldsymbol{u})$ is given by

$$\left[\mathrm{prox}_{\gamma\|\cdot\|_{\ell^1}}(\boldsymbol{u})\right]_i = [S_\gamma(\boldsymbol{u})]_i \triangleq \begin{cases} u_i - \gamma, & u_i \geq \gamma, \\ 0, & |u_i| < \gamma, \\ u_i + \gamma, & u_i \leq -\gamma, \end{cases} \tag{10.45}$$

where u_i is the i-th element of vector $\boldsymbol{u}$.

For the variables $\boldsymbol{z}_1$ and $\boldsymbol{z}_2$, we need to compute the proximal operators of indicator functions. From (4.38) (p. 73), the proximal operator of the indicator function on a closed and convex set $\mathcal{C}$ is given by the projection $\Pi_{\mathcal{C}}$ onto $\mathcal{C}$. Therefore, the second step for variables $\boldsymbol{z}_1$ and $\boldsymbol{z}_2$ are reduced to $\Pi_{\mathcal{C}_1}$ and $\Pi_{\mathcal{C}_2}$.

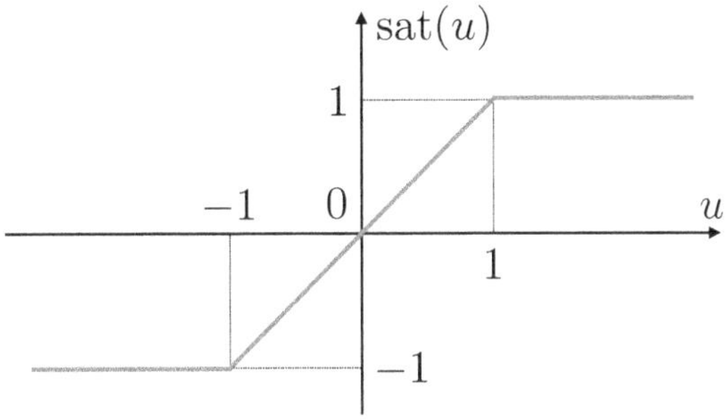

Figure 10.3: Saturation function $\mathrm{sat}(u) = \mathrm{sgn}(u)\min\{|u|, 1\}$.

The projection Π_{C_1} is given by

$$\Pi_{C_1}(\boldsymbol{u}) = \begin{bmatrix} \mathrm{sat}(u_1) \\ \mathrm{sat}(u_2) \\ \vdots \\ \mathrm{sat}(u_n) \end{bmatrix}, \quad \mathrm{sat}(u) \triangleq \mathrm{sgn}(u)\min\{|u|, 1\}, \tag{10.46}$$

where the function $\mathrm{sat}(\cdot)$ is called the *saturation function.* Figure 10.3 shows the graph of the saturation function. The other projection $\Pi_{C_2} = \Pi_{\{\zeta\}}$ is simply given by

$$\Pi_{C_2}(\boldsymbol{z}) \triangleq \boldsymbol{\zeta}. \tag{10.47}$$

In summary, the second step for variable $\boldsymbol{z}$ is given by

$$\boldsymbol{z}[k+1] = \begin{bmatrix} S_\gamma(\boldsymbol{u}[k+1] + \boldsymbol{v}_0[k]) \\ \Pi_{C_1}(\boldsymbol{u}[k+1] + \boldsymbol{v}_1[k]) \\ \boldsymbol{\zeta} \end{bmatrix}, \tag{10.48}$$

where we split the vector $\boldsymbol{v}[k]$ as $\boldsymbol{v} = [\boldsymbol{v}_0^\top, \boldsymbol{v}_1^\top, \boldsymbol{v}_2^\top]^\top$ consistent with the split of $\boldsymbol{z}$ in (10.35).

Now we obtain the ADMM algorithm to solve the ℓ^1 optimization (10.26):

ADMM algorithm to solve the ℓ^1 optimization problem (10.26)

Initialization: give initial vectors $\boldsymbol{z}[0]$, $\boldsymbol{v}[0] \in \mathbb{R}^\nu$, and real number $\gamma > 0$.

Iteration: for $k = 0, 1, 2, \dots$ do

$$\boldsymbol{u}[k+1] = M(\boldsymbol{z}[k] - \boldsymbol{v}[k]), \tag{10.49}$$

$$\boldsymbol{z}[k+1] = \begin{bmatrix} S_\gamma(\boldsymbol{u}[k+1] + \boldsymbol{v}_0[k]) \\ \Pi_{\mathcal{C}_1}(\boldsymbol{u}[k+1] + \boldsymbol{v}_1[k]) \\ \boldsymbol{\zeta} \end{bmatrix}, \tag{10.50}$$

$$\boldsymbol{v}[k+1] = \boldsymbol{v}[k] + \Psi \boldsymbol{u}[k+1] - \boldsymbol{z}[k+1], \quad k = 0, 1, 2, \dots \tag{10.51}$$

In this algorithm, the matrix M in (10.49) is given by (10.42). The Python implementation of the above algorithm is given in Section 10.6.2.

As mentioned in [13], the ADMM algorithm is very fast and requires just a few dozens of iterations to obtain a solution with sufficient precision. This property is very important if you adapt the finite-horizon L^1 optimal control to *model predictive control* [88], where real-time computation is essential.

10.5 Further Readings

The time discretization discussed in this section is based on the fundamental theory of sampled-data control, for which you can refer to a standard textbook by Chen and Francis [26]. The concept of pathological sampling is also found in this book.

10.6 Python Programs

We show Python programs to solve the ℓ^1 optimization problem (10.26). One is a program using CVXPY. The other is an implementation of the ADMM algorithm.

10.6.1 Python program based on CVXPY

The following program solves the ℓ^1 optimization problem in (10.26) via CVXPY.

```
import numpy as np
from numpy import linalg as LA
import cvxpy as cp
```

```python
import matplotlib.pyplot as plt
from scipy.signal import cont2discrete as c2d

# System model
A = np.array([[0, 1], [0, 0]])
b = np.array([[0], [1]])
d = len(b)  # system size
x0 = np.array([[1], [1]])  # initial states
T = 5  # Horizon length

# Time discretization
n = 1000  # grid size
h = T / n  # discretization interval
Ad,bd,_,_,_ = c2d((A,b,None,None), h) #c2d

# Matrix Phi
Phi = np.zeros((d, n))
v = bd
Phi[:, -1] = v.flatten()
for j in range(1, n):
    v = Ad @ v
    Phi[:, -j - 1] = v.flatten()

# Vector zeta
Ad_n = LA.matrix_power(Ad, n)
zeta = -Ad_n @ x0.flatten()

# Convex optimization via CVXPY
u = cp.Variable(n)
objective = cp.Minimize(cp.norm(u, 1))
constraints = [Phi @ u == zeta] + [cp.norm(u, np.
    inf) <= 1]
problem = cp.Problem(objective, constraints)
problem.solve()

# Plot
plt.figure()
plt.plot(np.arange(0, T, h), u.value)
```

```
plt.title('Sparse control')
plt.xlabel('Time')
plt.ylabel('Control input')
plt.show()
```

10.6.2 Python program based on ADMM

The following program solves the ℓ^1 optimization problem in (10.26) based on the ADMM algorithm.

```
import numpy as np
from numpy import linalg as LA
import matplotlib.pyplot as plt
from scipy.signal import cont2discrete as c2d

# System model
A = np.array([[0, 1], [0, 0]])
b = np.array([[0], [1]])
d = len(b)  # system size
x0 = np.array([[1], [1]])  # initial states
T = 5  # Horizon length

# Time discretization
n = 1000  # grid size
h = T / n  # discretization interval
Ad,bd,_,_,_ = c2d((A,b,None,None), h) #c2d

# Matrix Phi
Phi = np.zeros((d, n))
v = bd
Phi[:, -1] = v.flatten()
for j in range(1, n):
    v = Ad @ v
    Phi[:, -j - 1] = v.flatten()

# Vector zeta
Ad_n = LA.matrix_power(Ad, n)
zeta1 = -Ad_n @ x0.flatten()
```

```
zeta = zeta1.reshape(-1,1)

# ADMM parameters
mu = 2 * n + d
Psi = np.vstack((np.eye(n), np.eye(n), Phi))
PsiT = Psi.T
In = np.eye(n)
Id = np.eye(d)
M2 = 0.5*In - 0.5*Phi.T@ fcduikmLA.inv(2*Id +
    Phi@Phi.T)@Phi
EPS = 1e-4
MAX_ITER = 10000
z = np.concatenate((np.zeros((2 * n, 1)), zeta))
v = np.zeros((mu, 1))
r = zeta
k = 0
gamma = 0.05

# Soft-thresholding function
def soft_thresholding(gamma, x):
    return np.sign(x) * np.maximum(np.abs(x)-gamma
        ,0)

# Saturation function
sat = lambda x: np.sign(x) * np.minimum(np.abs(x),
    1)

# ADMM iterations
while (LA.norm(r) > EPS) and (k < MAX_ITER):
    u2 = PsiT @ (z-v)
    u = M2 @ u2
    z0 = soft_thresholding(gamma, u[:n] + v[:n])
    z1 = sat(u + v[n:n + n])
    z2 = zeta
    z = np.concatenate((z0, z1, z2))
    v = v + Psi @ u - z
    r = Phi @ u - zeta
    k = k + 1
```

```

# Plot
plt.figure()
plt.plot(np.arange(0, T, h), u, linewidth=2)
plt.title('Sparse control')
plt.xlabel('Time')
plt.ylabel('Control input')
plt.show()
```

Chapter 11

Advanced Topics

In this chapter, we introduce advanced topics in maximum hands-off control.

11.1 Smooth Hands-off Control by Mixed L^1/L^2 Optimization

As we studied in Chapter 9, the maximum hands-off control (the L^0-optimal control) is bang-off-bang (Theorem 9.6, p. 208). That is, the maximum hands-off control is a piecewise constant function taking values of ± 1 and 0. This means that the maximum hands-off control is *discontinuous*; the control changes its value between 1 and 0, or 0 and -1 at switching times. This is undesirable for some applications where actuators cannot move abruptly. In this case, one may want to make the control *continuous*. For this purpose, we add a regularization term to the L^1 cost $J_1(u)$ in the L^1 optimal control problem (Problem 9.2, p. 201). That is, we consider the following cost function:

$$J_{12}(u) = \lambda \|u\|_1 + \frac{1}{2}\|u\|_2^2 = \int_0^T \left(\lambda |u(t)| + \frac{1}{2}|u(t)|^2\right) dt, \tag{11.1}$$

where $\lambda > 0$ is a fixed parameter.

The idea of adding the L^2 norm term is borrowed from the *elastic net regularization*[1] in compressed sensing [158]. The elastic net regularization promotes sparsity with the grouping effect, where strongly correlated vectors are chosen at the same time. This ensures that the solution is not overly sensitive to small changes in the observation. From this idea, the L^2 term in (11.1) enhances the continuity of the solution.

[1]The name "elastic net" is meant to suggest a stretchable fishing net that retains all the big fish.

With the cost function (11.1), we consider the following mixed L^1/L^2-optimal control problem.

Problem 11.1 (L^1/L^2-optimal control problem). For the linear time-invariant system

$$\dot{\boldsymbol{x}}(t) = A\boldsymbol{x}(t) + \boldsymbol{b}u(t), \quad t \geq 0, \quad \boldsymbol{x}(0) = \boldsymbol{\xi} \in \mathbb{R}^d, \tag{11.2}$$

find a control $\{u(t) : t \in [0,T]\}$ with $T > 0$ that minimizes

$$J_{12}(u) = \lambda \|u\|_1 + \frac{1}{2}\|u\|_2^2, \tag{11.3}$$

subject to $\boldsymbol{x}(T) = \boldsymbol{0}$ and $\|u\|_\infty \leq 1$.

To discuss properties of the L^1/L^2-optimal control, we give necessary conditions of the optimality by Pontryagin's minimum principle.

The Hamiltonian function associated with Problem 11.1 is given by

$$H^\eta(\boldsymbol{x}, \boldsymbol{p}, u) = \boldsymbol{p}^\top (A\boldsymbol{x} + \boldsymbol{b}u) + \eta \left(\lambda |u| + \frac{1}{2}|u|^2 \right). \tag{11.4}$$

We do not consider the abnormal case (i.e., $\eta = 0$) and assume $\eta = 1$. Let $u^*(t)$ denote the optimal control and $\boldsymbol{x}^*(t)$ and $\boldsymbol{p}^*(t)$ the resultant optimal state and costate, respectively. Then, we have the following result.

Lemma 11.1. The L^1/L^2-optimal control $u^*(t)$ satisfies

$$u^*(t) = -\mathrm{sat}\Big(S_\lambda \left(\boldsymbol{p}^*(t)^\top \boldsymbol{b} \right) \Big), \tag{11.5}$$

where $S_\lambda(\cdot)$ is the soft-thresholding operator (see Section 4.2.5, p. 73) defined by

$$S_\lambda(v) \triangleq \begin{cases} v + \lambda & \text{if } v < -\lambda, \\ 0, & \text{if } -\lambda \leq v \leq \lambda, \\ v - \lambda, & \text{if } \lambda < v, \end{cases} \tag{11.6}$$

and $\mathrm{sat}(\cdot)$ is the saturation function defined by

$$\mathrm{sat}(v) \triangleq \begin{cases} -1, & \text{if } v < -1, \\ v, & \text{if } -1 \leq v \leq 1, \\ 1, & \text{if } 1 < v. \end{cases} \tag{11.7}$$

See Figure 11.1 for the graphs of $\mathrm{sat}(S_\lambda(v))$ in (11.5).

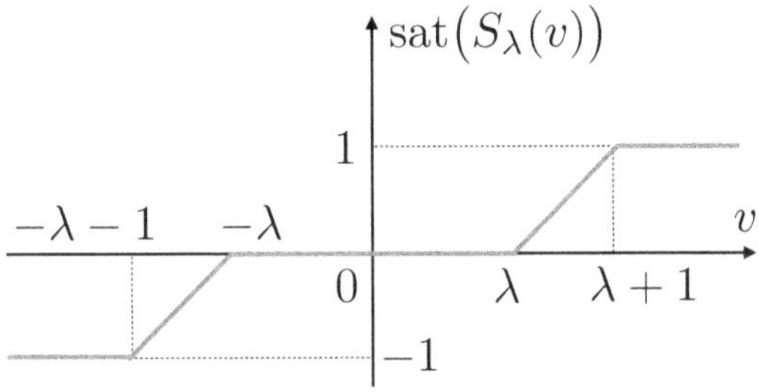

Figure 11.1: Saturated shrinkage function $\text{sat}(S_\lambda(v))$

Proof of Lemma 11.1: From Pontryagin's minimum principle, we have

$$\begin{aligned} u^*(t) &= \underset{u\in[-1,1]}{\arg\min} \left\{ (\boldsymbol{p}^*(t)^\top \boldsymbol{b})u + \lambda|u| + \frac{1}{2}|u|^2 \right\} \\ &= \begin{cases} 1, & \text{if } \boldsymbol{p}^*(t)^\top \boldsymbol{b} \le -\lambda - 1, \\ -(\boldsymbol{p}^*(t)^\top \boldsymbol{b} + \lambda), & \text{if } -\lambda - 1 < \boldsymbol{p}^*(t)^\top \boldsymbol{b} < -\lambda, \\ 0, & \text{if } -\lambda \le \boldsymbol{p}^*(t)^\top \boldsymbol{b} \le \lambda, \\ -(\boldsymbol{p}^*(t)^\top \boldsymbol{b} - \lambda), & \text{if } \lambda < \boldsymbol{p}^*(t)^\top \boldsymbol{b} < \lambda + 1, \\ -1, & \text{if } \lambda + 1 \le \boldsymbol{p}^*(t)^\top \boldsymbol{b}, \end{cases} \\ &= -\text{sat}\Big(S_\lambda\left(\boldsymbol{p}^*(t)^\top \boldsymbol{b}\right)\Big). \end{aligned} \tag{11.8}$$

□

From Lemma 11.1, we have the following theorem.

Theorem 11.1 (Continuity). The L^1/L^2-optimal control $u^*(t)$ is continuous in t over $[0, T]$.

Proof: Define

$$\bar{u}(\boldsymbol{p}) \triangleq -\text{sat}\Big(S_\lambda\left(\boldsymbol{p}^\top \boldsymbol{b}\right)\Big). \tag{11.9}$$

Since the composite function $\text{sat} \circ S_\lambda$ is continuous (see Figure 11.1), $\bar{u}(\boldsymbol{p})$ is also continuous in $\boldsymbol{p}$. It follows from Lemma 11.1 that the optimal control u^* given in (11.5) is continuous in $\boldsymbol{p}^*$. Hence, $u^*(t)$ is continuous, if $\boldsymbol{p}^*(t)$ is continuous in t over $[0, T]$. In fact, from (9.26) (p. 203), $\boldsymbol{p}^*(t)^\top \boldsymbol{b}$ is given by

$$\boldsymbol{p}^*(t)^\top \boldsymbol{b} = \boldsymbol{p}^*(0)^\top e^{-At} \boldsymbol{b}, \tag{11.10}$$

which is continuous in t over $\mathbb{R}$. □

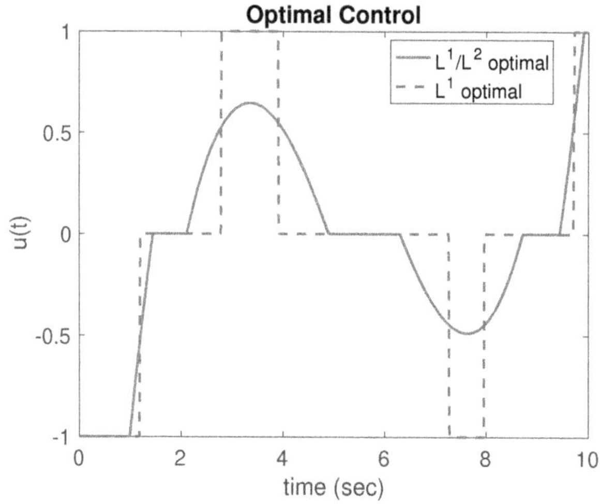

Figure 11.2: Maximum hands-off control (dashed) and L^1/L^2-optimal control (solid)

Theorem 11.1 motivates us to use the L^1/L^2-optimal control for continuous hands-off control. In general, the degree of continuity (or smoothness) and the sparsity of the control input cannot be optimized at the same time. The parameter λ can be used for trading smoothness for sparsity. Lemma 11.1 suggests that increasing the parameter λ makes the L^1/L^2 optimal control $u^*(t)$ sparser (see also Fig. 11.1). On the other hand, decreasing λ smoothens $u^*(t)$.

Example 11.1. Let us consider the following linear system

$$\frac{d\boldsymbol{x}(t)}{dt} = \begin{bmatrix} 0 & 1 & 0 & 0 \\ 0 & 0 & 1 & 0 \\ 0 & 0 & 0 & 1 \\ 0 & 0 & 0 & 0 \end{bmatrix} \boldsymbol{x}(t) + \begin{bmatrix} 0 \\ 0 \\ 0 \\ 1 \end{bmatrix} u(t). \tag{11.11}$$

We set the final time $T = 10$, and the initial and final states as

$$\boldsymbol{x}(0) = [0.5, 0.5, 0.5, 0.5]^\top, \quad \boldsymbol{x}(10) = \boldsymbol{0}. \tag{11.12}$$

Fig. 11.2 shows the L^1/L^2 optimal control with weights $\lambda = 1$. The maximum hands-off control is also illustrated. We can see that the L^1/L^2-optimal control is continuous but sufficiently sparse. □

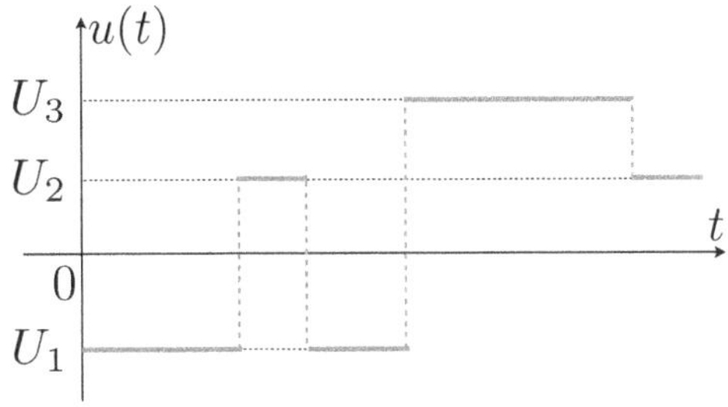

Figure 11.3: An example of discrete-valued control that takes three values of U_1, U_2, and U_3.

11.2 Discrete-valued Control

As we observed in Chapter 9, the maximum hands-off control (or the L^0-optimal control) takes values in an *alphabet*[2] $\{-1, 0, 1\}$. Such a control is called a *discrete-valued control*, since the control takes a finite number of values. Discrete-valued control is important in networked control systems where the bandwidth of the network is limited, since discrete-valued signals can be effectively compressed.

We here generalize the property of discreteness in maximum hands-off control by the sum-of-absolute-values (SOAV) optimization.

11.2.1 Sum-of-absolute-values (SOAV) optimization

Let us consider discrete-valued control for the linear time-invariant plant

$$\dot{\boldsymbol{x}}(t) = A\boldsymbol{x}(t) + \boldsymbol{b}u(t), \quad t \geq 0, \quad \boldsymbol{x}(0) = \boldsymbol{\xi} \in \mathbb{R}^d, \tag{11.13}$$

where the control $u(t)$ takes N real numbers

$$U_1 < U_2 < \cdots < U_N. \tag{11.14}$$

That is, we consider a discrete-valued control with alphabet $\{U_1, U_2, \ldots, U_N\}$. Figure 11.3 shows an example of discrete-valued control. The purpose we consider here is to seek a discrete-valued control that achieves $\boldsymbol{x}(T) = \boldsymbol{0}$, given the initial state $\boldsymbol{x}(0) = \boldsymbol{\xi}$ and the control time $T > 0$.

A standard method to obtain discrete-valued control is to describe the problem as a mixed-integer programming problem [8]. However, this method requires a lot of computational time, which grows exponentially as the size of the problem grows, and hence this method is hard to apply

[2]The word *alphabet* is borrowed from information theory [31]. An alphabet is a set of a finite number of elements that are used to represent signals of interest.

to a large-scale problem. Instead, we consider convex relaxation of this optimization problem of discrete-valued control.

We first define the *feasible controls* that drive the state $\boldsymbol{x}(t)$ from the initial state $\boldsymbol{x}(0) = \boldsymbol{\xi}$ to the origin in time $T > 0$, satisfying

$$U_1 \leq u(t) \leq U_N, \quad \forall t \in [0, T]. \tag{11.15}$$

We denote by $\mathcal{U}(T, \boldsymbol{\xi})$ the set of feasible controls. We assume that $\boldsymbol{\xi} \in \mathbb{R}^d$ and $T > 0$ are given such that $\mathcal{U}(T, \boldsymbol{\xi})$ is non-empty. For a feasible control $u \in \mathcal{U}(T, \boldsymbol{\xi})$, define the following cost function:

$$J_0(u) \triangleq \sum_{j=1}^{N} w_j \|u - U_j\|_0, \tag{11.16}$$

where $w_1, w_2, \ldots, w_N$ are weights that satisfy

$$w_i > 0, \quad w_1 + w_2 + \cdots + w_N = 1. \tag{11.17}$$

Minimizing the cost function (11.16) may promote discreteness of the control to take values in $\{U_1, \ldots, U_N\}$. This can be explained as follows. A discrete-valued control is a piecewise constant signal as shown in Figure 11.3. If $u(t) = U_j$ for t in some time intervals with a positive length, then the function $u(t) - U_j$ is zero over the intervals, and hence it is *sparse*. Namely, the L^0 norm of the function $u - U_j$ should be smaller than T. If we choose the weights $w_1, \ldots, w_N$ according to the importance of the values $U_1, \ldots, U_N$ and minimize the cost function (11.16), we may obtain a discrete-valued feasible control.

The cost function (11.16) is discontinuous and non-convex, and hence it is difficult to directly obtain the optimal solution as in the case of L^0-optimal control. We then adopt the L^1 relaxation, that is, we use the L^1 norm instead of the L^0 norm in (11.16):

$$J_1(u) \triangleq \sum_{j=1}^{N} w_j \|u - U_j\|_1 = \int_0^T \sum_{j=1}^{N} w_j |u(t) - U_j| dt. \tag{11.18}$$

We call this cost function the *sum of absolute values* or *SOAV* for short. Then, we describe the SOAV-optimal control problem as follows:

Problem 11.2 (SOAV-optimal control problem). For the linear time-invariant system

$$\dot{\boldsymbol{x}}(t) = A\boldsymbol{x}(t) + \boldsymbol{b}u(t), \quad t \geq 0, \quad \boldsymbol{x}(0) = \boldsymbol{\xi} \in \mathbb{R}^d, \tag{11.19}$$

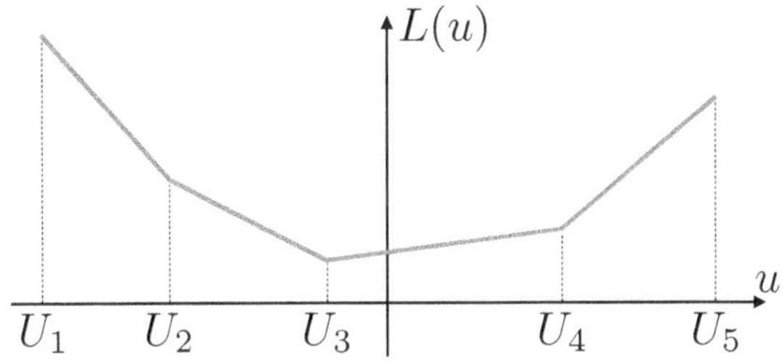

Figure 11.4: Piecewise linear function $L(u)$

find a control $\{u(t) : t \in [0, T]\}$ that minimizes

$$J_1(u) = \sum_{j=1}^{N} w_j \|u - U_j\|_1 = \int_0^T \sum_{j=1}^{N} w_j |u(t) - U_j| dt, \tag{11.20}$$

subject to $\boldsymbol{x}(T) = \mathbf{0}$ and $U_1 \leq u(t) \leq U_N$ for all $t \in [0, T]$.

The optimal control is called the *sum-of-absolute-values optimal control* or *SOAV-optimal control.*

11.2.2 Discreteness of SOAV-optimal control

Here we show that the SOAV-optimal control is a discrete-valued control taking values in $\{U_1, \ldots, U_N\}$ under some conditions.

Let $u^* \in \mathcal{U}(T, \boldsymbol{\xi})$ be an SOAV-optimal control minimizing the cost function (11.18), that is,

$$u^* = \arg\min_u \; J_1(u) \; \text{ subject to } \; u \in \mathcal{U}(T, \boldsymbol{\xi}). \tag{11.21}$$

For the optimal control problem (Problem 11.2), we analyze the solution u^* by using Pontryagin's minimum principle.

The stage cost function $L(u)$ of the SOAV cost function (11.18) is given by

$$L(u) = \sum_{j=1}^{N} w_j |u - U_j|. \tag{11.22}$$

Figure 11.4 shows an example of function $L(u)$. As shown in this figure, the stage cost function $L(u)$ is a continuous and piecewise linear function. Also, since the function $L(u)$ is a convex combination of convex functions $|u - U_j|$, $j = 1, \ldots, N$, $L(u)$ is convex in u. That is, the optimization problem in Problem 11.2 is a convex optimization problem.

Then the Hamiltonian for Problem 11.2 is defined by

$$\begin{aligned} H^{\eta}(\boldsymbol{x}, \boldsymbol{p}, u) &= \boldsymbol{p}^{\top}(A\boldsymbol{x} + \boldsymbol{b}u) + \eta L(u) \\ &= \boldsymbol{p}^{\top}(A\boldsymbol{x} + \boldsymbol{b}u) + \eta \sum_{j=1}^{N} w_j |u - U_j|. \end{aligned} \tag{11.23}$$

Here we assume $\eta = 1$. Let $\boldsymbol{x}^*$ and $\boldsymbol{p}^*$ be respectively the optimal state and costate with the optimal control u^*. From the minimum principle, we have

$$\begin{aligned} u^*(t) &= \underset{u \in [U_1, U_N]}{\arg\min} \ \{\boldsymbol{p}^*(t)^{\top}(A\boldsymbol{x}^*(t) + \boldsymbol{b}u) + L(u)\} \\ &= \underset{u \in [U_1, U_N]}{\arg\min} \ \{\boldsymbol{p}^*(t)^{\top}\boldsymbol{b}u + L(u)\}. \end{aligned} \tag{11.24}$$

Let us solve the minimization problem in (11.24).

Since the function $L(u)$ is piecewise linear, $L(u)$ can be written as

$$L(u) = \begin{cases} a_1 u + b_1, & u \in [U_1, U_2], \\ a_2 u + b_2, & u \in [U_2, U_3], \\ & \vdots \\ a_{N-1} u + b_{N-1}, & u \in [U_{N-1}, U_N], \end{cases} \tag{11.25}$$

where

$$\begin{aligned} a_k &= \sum_{j=1}^{k} w_j - \sum_{j=k+1}^{N} w_j, \\ b_k &= -\sum_{j=1}^{k} w_j U_j + \sum_{j=k+1}^{N} w_j U_j, \quad k = 1, 2, \ldots, N-1. \end{aligned} \tag{11.26}$$

Fix $t \in [0, T]$ and define $\alpha \triangleq \boldsymbol{p}^*(t)^{\top}\boldsymbol{b} \in \mathbb{R}$. Since $L(u)$ is continuous, and the following inequality

$$a_1 < a_2 < \cdots < a_{N-1} \tag{11.27}$$

holds, we can compute the minimizer of

$$h(u) \triangleq \alpha u + L(u) = \begin{cases} (a_1 + \alpha) u + b_1, & u \in [U_1, U_2], \\ (a_2 + \alpha) u + b_2, & u \in [U_2, U_3], \\ & \vdots \\ (a_{N-1} + \alpha) u + b_{N-1}, & u \in [U_{N-1}, U_N], \end{cases} \tag{11.28}$$

for $u \in [U_1, U_N]$.

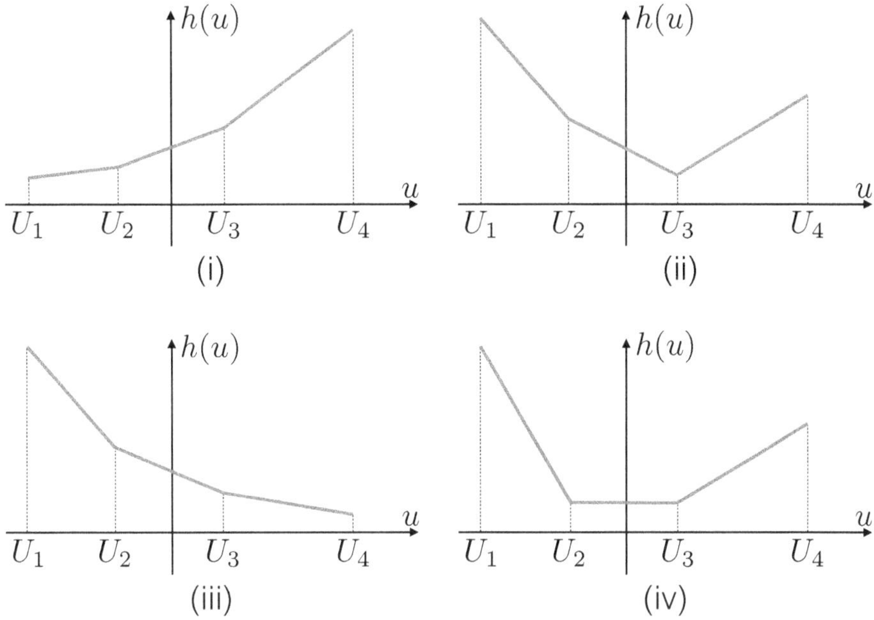

Figure 11.5: 4 cases of piecewise linear function $h(u) = \alpha u + L(u)$

(i) If $a_1 + \alpha > 0$, then from (11.27) we have

$$0 < a_1 + \alpha < a_2 + \alpha < \cdots < a_{N-1} + \alpha, \tag{11.29}$$

and the slopes $(a_k + \alpha)$ of the linear functions in (11.28) are all positive. See (i) of Figure 11.5. Hence we have

$$\underset{u \in [U_1, U_N]}{\arg\min}\ h(u) = U_1. \tag{11.30}$$

(ii) If $a_k + \alpha < 0$ and $a_{k+1} + \alpha > 0$ $(k = 1, \ldots, N-2)$, then from (11.27) we have

$$a_1 + \alpha < a_2 + \alpha < \cdots < a_k + \alpha < 0, \tag{11.31}$$

and

$$0 < a_{k+1} + \alpha < a_{k+2} + \alpha < \cdots < a_{N-1} + \alpha. \tag{11.32}$$

The sign of the slopes of the linear functions in (11.28) changes from negative to positive at $u = U_{k+1}$. See (ii) of Figure 11.5. Hence, we have

$$\underset{u \in [U_1, U_N]}{\arg\min}\ h(u) = U_{k+1}. \tag{11.33}$$

(iii) If $a_{N-1} + \alpha < 0$, then we have

$$a_1 + \alpha < a_2 + \alpha < \cdots < a_{N-1} + \alpha < 0, \tag{11.34}$$

and the slopes $(a_k + \alpha)$ in (11.28) are all negative. See (iii) of Figure 11.5. Hence we have

$$\arg\min_{u \in [U_1, U_N]} h(u) = U_N. \tag{11.35}$$

(iv) If there exists $k \in \{1, 2, \ldots, N-1\}$ such that $a_k + \alpha = 0$, then the slope becomes zero over the interval $[U_k, U_{k+1}]$. Hence we have

$$\arg\min_{u \in [U_1, U_N]} h(u) = [U_k, U_{k+1}]. \tag{11.36}$$

In this case, we cannot determine the unique value for $u^*(t)$.

In summary, the SOAV-optimal control $u^*(t)$ satisfies the following:

$$u^*(t) = \begin{cases} U_1, & \text{if } -a_1 < \boldsymbol{p}^*(t)^\top \boldsymbol{b}, \\ U_2, & \text{if } -a_2 < \boldsymbol{p}^*(t)^\top \boldsymbol{b} < -a_1, \\ & \vdots \\ U_{N-1}, & \text{if } -a_{N-1} < \boldsymbol{p}^*(t)^\top \boldsymbol{b} < -a_{N-2}, \\ U_N, & \text{if } \boldsymbol{p}^*(t)^\top \boldsymbol{b} < -a_{N-1}, \end{cases} \tag{11.37}$$

and

$$u^*(t) \in [U_k, U_{k+1}], \quad \text{if } \boldsymbol{p}^*(t)^\top \boldsymbol{b} = -a_k, \quad k = 1, 2, \ldots, N-1. \tag{11.38}$$

From (11.38), if

$$\boldsymbol{p}^*(t)^\top \boldsymbol{b} \neq -a_k, \quad k = 1, 2, \ldots, N-1, \tag{11.39}$$

holds for almost all $t \in [0, T]$, then we can see that $u^*(t)$ takes discrete-values in $\{U_1, \ldots, U_N\}$ for almost all $t \in [0, T]$. Let us consider a sufficient condition for this.

We see that (11.39) holds for almost all $t \in [0, T]$ if and only if

$$\mu(\{t \in [0, T] : \boldsymbol{p}^*(t)^\top \boldsymbol{b} = -a_k\}) = 0 \tag{11.40}$$

for $k = 1, 2, \ldots, N-1$. We say the SOAV-optimal control is *non-singular* if (11.40) holds for all $k \in \{1, 2, \ldots, N-1\}$. Then we have the following theorem:

Theorem 11.2. Assume that the SOAV-optimal control is non-singular. Then the optimal control $u^*(t)$ takes values in $\{U_1, \ldots, U_N\}$ for almost all $t \in [0, T]$.

For the non-singularity, we have the following theorem.

Theorem 11.3. Assume that the pair $(A, \boldsymbol{b})$ is non-singular. That is, the pair $(A, \boldsymbol{b})$ is controllable and A is non-singular. Assume also that

$$\sum_{j=1}^{k} w_j \neq \sum_{j=k+1}^{N} w_j \tag{11.41}$$

holds for $k = 1, 2, \ldots, N-1$. Then the SOAV-optimal control is non-singular.

Exercise 11.1. Prove Theorem 11.3.

The condition (11.41) in Theorem 11.3 is a sufficient and necessary condition for the slopes of the linear functions in (11.28) to be nonzero.

Example 11.2. Let us consider a design example of SOAV-optimal control. We consider the 4-th order plant given in (11.11) in Example 11.1. The final time $T = 10$ and the initial and final states are the same as (11.12).

The alphabet is given by $\{-1, -0.5, 0, 0.5, 1\}$, that is, $N = 5$ and

$$U_1 = -1, \ U_2 = -0.5, \ U_3 = 0, \ U_4 = 0.5, \ U_5 = 1. \tag{11.42}$$

The weights in the cost function (11.18) are set as

$$w_1 = w_2 = w_3 = w_4 = w_5 = \frac{1}{5}. \tag{11.43}$$

□

Figure 11.6 shows the obtained SOAV-optimal control. In this figure, the L^1-optimal control (or the maximum hands-off control) discussed in Chapter 9 and the L^2-optimal control[3] that minimizes the L^2 norm

$$J_2(u) = \int_0^T |u(t)|^2 dt, \tag{11.44}$$

are also shown. Note that the L^1-optimal control is bang-off-bang and takes values of ± 1 and 0. On the other hand, the L^2-optimal control is

[3] The L^2-optimal control is also known as *minimum-energy control* [3, Section 6-18].

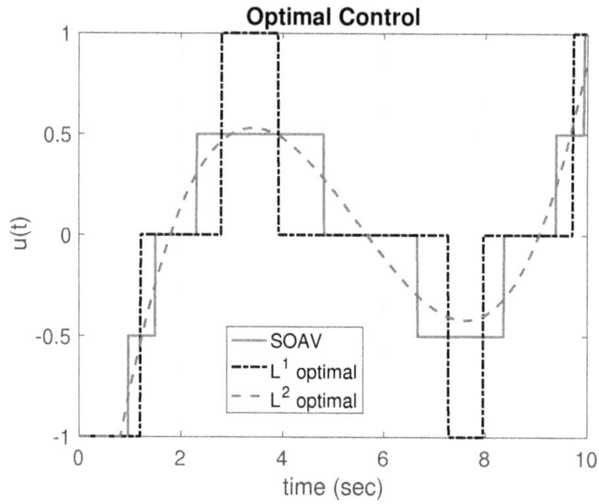

Figure 11.6: SOAV-optimal control (solid), L^1-optimal control (dashed), and L^2-optimal control (dotted).

a smooth control. The SOAV-optimal control is between them. It takes discrete values in the alphabet $\{-1, -0.5, 0, 0.5, 1\}$, that is a quantization of the L^2-optimal control.

Figure 11.7 shows the state variables $x_1(t), \ldots, x_4(t)$ in the state $\boldsymbol{x}(t)$ and the SOAV-optimal control. We can see that by the obtained discrete-valued control $u(t)$, all the state variables converge to the origin in the time $T = 10$. Note that this cannot be possible when one uses a quantized version of the L^2-optimal control by a static quantizer; there should be quantization errors that perturb the state trajectory.

11.3 Time-optimal Hands-off Control

In this section, we consider an optimal control that takes account of sparsity and time-optimality at the same time. Let us consider the following linear time-invariant system:

$$\dot{\boldsymbol{x}}(t) = A\boldsymbol{x}(t) + \boldsymbol{b}u(t), \quad t \geq 0, \quad \boldsymbol{x}(0) = \boldsymbol{\xi} \in \mathbb{R}^d. \tag{11.45}$$

The control objective is to drive the state to the origin. Here we do not fix the final time T. As in the minimum-time control in Chapter 8, the final time T is also an optimization variable.

First, we consider the feasibility of the control. For the system (11.45), a control u is said to be feasible if there exists a finite time $T > 0$ such

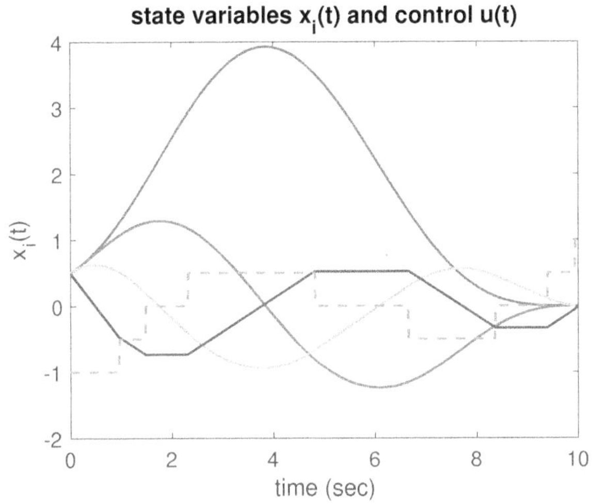

Figure 11.7: State variables $x_1(t), \ldots, x_4(t)$ and the SOAV-optimal control $u(t)$, $t \in [0, 10]$.

that by $\{u(t) : t \in [0, T]\}$ satisfying

$$|u(t)| \leq 1, \quad \forall t \in [0, T], \tag{11.46}$$

the state $\boldsymbol{x}(t)$ in (11.45) is steered from $\boldsymbol{x}(0) = \boldsymbol{\xi}$ to $\boldsymbol{x}(T) = \mathbf{0}$. From the definition of the controllable set $\mathcal{R}$ in (8.30) (p. 183), there exists a feasible control if the initial state $\boldsymbol{\xi}$ is in the controllable set $\mathcal{R}$. Therefore, we assume $\boldsymbol{\xi} \in \mathcal{R}$. Using the feasible set $\mathcal{U}(T, \boldsymbol{\xi})$ with fixed $T > 0$ (see Section 8.1.3), the set of all feasible controls is given by

$$\mathcal{U}(\boldsymbol{\xi}) \triangleq \bigcup_{T \geq 0} \mathcal{U}(T, \boldsymbol{\xi}). \tag{11.47}$$

Next, we formulate the optimal control problem. We seek a feasible control $u \in \mathcal{U}(\boldsymbol{\xi})$ that minimizes the L^0 norm of u and the response time T at the same time. For this, we consider the following cost function:

$$J_0(u) \triangleq \lambda \|u\|_0 + T, \tag{11.48}$$

where $\lambda > 0$ is a parameter for a tradeoff between the two requirements. As usual, we relax the L^0 norm in (11.48) by the L^1 norm $\|u\|_1$, namely, we consider the following cost function:

$$J_1(u) \triangleq \lambda \|u\|_1 + T. \tag{11.49}$$

Now we formulate our problem.

Problem 11.3 (L^1-time-optimal control problem). For the linear time-invariant system

$$\dot{\boldsymbol{x}}(t) = A\boldsymbol{x}(t) + \boldsymbol{b}u(t), \quad t \geq 0, \quad \boldsymbol{x}(0) = \boldsymbol{\xi} \in \mathbb{R}^d, \tag{11.50}$$

find a control $\{u(t) : t \in [0, \infty)\}$ that minimizes

$$J_1(u) = \lambda \|u\|_1 + T, \tag{11.51}$$

subject to $\boldsymbol{x}(T) = \boldsymbol{0}$ and $\|u\|_\infty \leq 1$.

We call the optimal solution the L^1*-time-optimal control.*

The existence theorem for the L^1-time-optimal control is proved similarly to the time-optimal control (Theorem 9.7, p. 209).

Theorem 11.4. For any initial state $\boldsymbol{\xi} \in \mathcal{R}$, there exists at least one L^1-time-optimal control.

You can find the proof in [66].

The Hamiltonian for Problem 11.3 is given by

$$H^\eta(\boldsymbol{x}, \boldsymbol{p}, u) = \boldsymbol{p}^\top (A\boldsymbol{x} + \boldsymbol{b}u) + \eta(\lambda|u| + 1). \tag{11.52}$$

We do not consider the abnormal case ($\eta = 0$) and assume $\eta = 1$. Then the optimal control $u^*(t)$ for Problem 11.3 satisfies

$$u^*(t) = \underset{u \in [-1,1]}{\arg\min}\, H^1(\boldsymbol{x}, \boldsymbol{p}, u) = \underset{u \in [-1,1]}{\arg\min} \left\{ \boldsymbol{p}^*(t)^\top \boldsymbol{b}u + \lambda|u| \right\}. \tag{11.53}$$

From this, we have

$$\begin{aligned} u^*(t) &= \begin{cases} 1, & \text{if } \ \boldsymbol{p}^*(t)^\top \boldsymbol{b} < -\lambda, \\ 0, & \text{if } \ -\lambda < \boldsymbol{p}^*(t)^\top \boldsymbol{b} < \lambda, \\ -1, & \text{if } \ \lambda < \boldsymbol{p}^*(t)^\top \boldsymbol{b}, \end{cases} \\ u^*(t) &\in [0, 1], \ \text{ if } \ \boldsymbol{p}^*(t)^\top \boldsymbol{b} = -\lambda, \\ u^*(t) &\in [-1, 0], \ \text{ if } \ \boldsymbol{p}^*(t)^\top \boldsymbol{b} = \lambda. \end{aligned} \tag{11.54}$$

If $\boldsymbol{p}^*(t)^\top \boldsymbol{b} = \pm\lambda$ holds only on sets of measure zero (i.e., if $\boldsymbol{p}^*(t)^\top \boldsymbol{b} \neq \pm\lambda$ almost all $t \in [0, T]$), then the L^1-time-optimal control is bang-off-bang. From Lemma 9.1, we have the following theorem:

Theorem 11.5. Assume that the pair $(A, \boldsymbol{b})$ is non-singular. Then the L^1-time-optimal control is bang-off-bang (if it exists).

This theorem, along with Theorem 11.4, leads to the following equivalence theorem:

Theorem 11.6. Assume $\boldsymbol{\xi} \in \mathcal{R}$ and the pair $(A, \boldsymbol{b})$ is non-singular. Then the L^1-time-optimal control is equivalent to the L^0-time-optimal control that minimizes the cost function (11.48).

11.4 Distributed Hands-off Control

In this section, we introduce *distributed* hands-off control over a network. As discussed in Section 6.2 (p. 128), consensus can be achieved by local averaging control. The local control is in general not sparse, and we can adapt the idea of maximum hands-off control to the consensus control.

Let us consider an undirected graph $G = (\mathcal{V}, \mathcal{E})$ with $\mathcal{V} = \{1, 2, \dots, N\}$. We assume G is connected. The i-th agent ($i \in \mathcal{V}$) has the following dynamics:

$$\dot{x}_i(t) = u_i(t), \quad i \in \mathcal{V}, \quad t \geq 0, \tag{11.55}$$

where we consider a 1-dimensional state $x_i(t) \in \mathbb{R}$ and a single input $u_i(t) \in \mathbb{R}$. We here assume a *sampled-data control* where the i-th agent can obtain the sampled data of states. More precisely, the agent i can get sampled-data $x_i(kT)$ and $x_j(kT)$, $j \in \mathcal{N}_i$ (the set of all neighbors of agent i), $k = 0, 1, 2, \dots$, with a given sampling period $T > 0$.

Now we formulate the problem of distributed hands-off control.

Problem 11.4 (Distributed hands-off control problem)**.** Find a control $\{u_i(t) : t \geq 0\}$ for agent $i \in \mathcal{V}$ that satisfies the following:

1. $\lim_{t\to\infty} |x_i(t) - x_j(t)| = 0$ for all $i, j \in \mathcal{V}$.
2. $|u(t)| \leq 1$ for all $i \in \mathcal{V}$ and all $t \geq 0$.
3. The local control $u_i(t)$, $t \in \mathcal{I}_k \triangleq [kT, (k+1)T)$ is determined by sampled states $x_i(kT)$ and $k_j(kT)$, $j \in \mathcal{N}_i$.

Adapting the consensus algorithm discussed in Section 6.2, we have the following local control of agent $i \in \mathcal{V}$ over k-th time interval $\mathcal{I}_k$:

$$u_i(t) = \arg\min \left\{ \|u\|_0 : u \in \mathcal{U}(T, x_i(kT), x_i^f[k]) \right\}, \quad t \in \mathcal{I}_k, \tag{11.56}$$

where $x_i^f[k]$ is the local final state defined by

$$x_i^f[k] \triangleq x_i(kT) - \epsilon \sum_{j \in \mathcal{N}_i} (x_i(kT) - x_j(kT)), \tag{11.57}$$

and $\mathcal{U}(T, \xi, \zeta)$ is the feasible set for the system (11.55) from $\xi \in \mathbb{R}$ to $\zeta \in \mathbb{R}$ by control u with $\|u\|_\infty \le 1$. Namely,

$$\mathcal{U}(T, \xi, \zeta) = \left\{ u \in L^\infty(0, T) : \zeta = \xi + \int_0^T u(t)dt, \ \|u\|_\infty \le 1 \right\}. \tag{11.58}$$

Since the local control $\{u_i(t) : t \in \mathcal{I}_k\}$ minimizes the L^0 norm, it is expected to be sparse. Moreover, the distributed control system achieves average consensus as described below.

Let L be the graph Laplacian of the graph G, and Δ be the maximum degree of G. Define the state vector $\boldsymbol{x}(t)$ by

$$\boldsymbol{x}(t) \triangleq \begin{bmatrix} x_1(t) \\ x_2(t) \\ \vdots \\ x_N(t) \end{bmatrix}. \tag{11.59}$$

Then we have the following theorem [60]:

Theorem 11.7. Assume G is connected. Assume also that the gain ϵ in (11.56) satisfies

$$0 < \epsilon\Delta < 1, \tag{11.60}$$

and the initial state vector $\boldsymbol{x}(0)$ satisfies

$$-T\mathbf{1}_N \le \epsilon L\boldsymbol{x}(0) \le T\mathbf{1}_N, \tag{11.61}$$

where $\mathbf{1}_N = \begin{bmatrix} 1 & 1 & \dots & 1 \end{bmatrix}^\top \in \mathbb{R}^N$. Then, there exists the local control $u_i(t)$ defined in (11.56), $t \in \mathcal{I}_k$ for any $i \in \mathcal{V}$ and any $k \in \{0, 1, 2, \dots\}$, and all states $x_i(t)$ converge to the average of the initial states:

$$\alpha \triangleq \frac{1}{N}\sum_{i=1}^{N} x_i(0). \tag{11.62}$$

11.5 Further Readings

The smooth hands-off control by the mixed L^1/L^2 optimization was first proposed in [105]. Another formulation for smooth hands-off control by the CLOT (Combined L-One and Two) norm was also proposed in [110]. The CLOT norm is defined by

$$\|u\|_{\mathrm{CLOT}} \triangleq \lambda_1\|u\|_1 + \lambda_2\|u\|_2, \tag{11.63}$$

with parameters $\lambda_1 > 0$ and $\lambda_2 > 0$ such that $\lambda_1 + \lambda_2 = 1$. Compared with the mixed L^1/L^2 cost function in (11.1), the L^2 term in the CLOT norm is not squared. The CLOT-optimal control is also continuous but sparser than the mixed L^1/L^2-optimal control in Problem 11.1.

The SOAV-optimal control has been proposed in [61], [65]. The idea of the SOAV cost function was first proposed in [97] for discrete-valued signal reconstruction. The SOAV optimization was then applied to digital communications [53], [135], [136].

The L^1-time-optimal control was first proposed in [66] and extended to L^1/ℓ^1-time-optimal control for sparsity in both time and space domain [68].

The theory of distributed hands-off control was established in [60], and then it was applied to distributed drone control in [95].

References

[1] M. Aldridge, L. Baldassini, and O. Johnson, "Group testing algorithms: Bounds and simulations", *IEEE Trans. Inf. Theory*, vol. 60, no. 6, Jun. 2014, pp. 3671–3687.

[2] R. Amirifar and N. Sadati, "Low-order H_∞ controller design for an active suspension system via LMIs", *IEEE Trans. Ind. Electron.*, vol. 53, no. 2, Apr. 2006.

[3] M. Athans and P. L. Falb, *Optimal Control.* Dover Publications, 2007.

[4] G. K. Atia and V. Saligrama, "Boolean compressed sensing and noisy group testing", *IEEE Trans. Inf. Theory*, vol. 58, no. 3, Mar. 2012, pp. 1880–1901.

[5] R. B. Bapat, *Graphs and Matrices.* Splinger, 2014.

[6] H. H. Bauschke and P. L. Combettes, *Convex Analysis and Monotone Operator Theory in Hilbert Spaces.* Springer, 2011.

[7] A. Beck and M. Teboulle, "Gradient-based algorithms with applications to signal-recovery problems", in *Convex Optimization*, Cambridge University Press, 2010.

[8] A. Bemporad and M. Morari, "Control of systems integrating logic, dynamics, and constraints", *Automatica*, vol. 35, 1999, pp. 407–427.

[9] D. Bertsekas, *Convex Optimization Algorithms.* Athena Scientific, 2015.

[10] C. M. Bishop, *Pattern Recognition and Machine Learning.* Springer, 2006.

[11] T. Blumensath and M. E. Davies, "Iterative thresholding for sparse approximations", *Journal of Fourier Analysis and Applications*, vol. 14, no. 5, 2008, pp. 629–654.

[12] S. Boyd, L. E. Ghaoui, E. Feron, and V. Balakrishnan, *Linear Matrix Inequalities in System and Control Theory.* SIAM, 1994.

[13] S. Boyd, N. Parikh, E. Chu, B. Peleato, and J. Eckstein, "Distributed optimization and statistical learning via the alternating direction method of multipliers", *Foundations and Trends in Machine Learning*, vol. 3, no. 1, 2011, pp. 1–122.

[14] S. Boyd and L. Vandenberghe, *Convex Optimization.* Cambridge University Press, 2004.

[15] J. P. Boyle and R. L. Dykstra, "A method for finding projections onto the intersection of convex sets in Hilbert spaces", in *Advances in Order Restricted Statistical Inference, Lecture Notes in Statistics*, R. Dykstra, T. Robertson, and F. T. Wright, Eds., vol. 37, New York: Springer, 1986.

[16] P. Bühlmann and S. van de Geer, *Statistics for High-Dimensional Data.* Springer, 2011.

[17] F. Bullo, *Lectures on Network Systems*, 1.7. Kindle Direct Publishing, 2024.

[18] K. Cai and M. Nagahara, "A new perspective on cooperative control of multi-agent systems through different types of graph laplacians", *Advanced Robotics*, vol. 37, no. 1-2, 2023, pp. 2–11.

[19] E. J. Candes and T. Tao, "Near-optimal signal recovery from random projections: Universal encoding strategies?", *IEEE Trans. Inf. Theory*, vol. 52, no. 12, Dec. 2006, pp. 5406–5425.

[20] Y. Cao, W. Yu, W. Ren, and G. Chen, "An overview of recent progress in the study of distributed multi-agent coordination", *IEEE Transactions on Industrial informatics*, vol. 9, no. 1, 2012, pp. 427–438.

[21] C. Chan, "The state of the art of electric, hybrid, and fuel cell vehicles", *Proc. IEEE*, vol. 95, no. 4, Apr. 2007, pp. 704–718.

[22] C. Chang and S. Sim, "Optimising train movements through coast control using genetic algorithms", *IEE Proceedings-Electric Power Applications*, vol. 144, no. 1, 1997, pp. 65–73.

[23] D. Chatterjee, M. Nagahara, D. E. Quevedo, and K. M. Rao, "Characterization of maximum hands-off control", *Systems & Control Letters*, vol. 94, 2016, pp. 31–36.

[24] S. S. Chen, D. L. Donoho, and M. A. Saunders, "Atomic decomposition by basis pursuit", *SIAM J. Sci. Comput.*, vol. 20, no. 1, Aug. 1998, pp. 33–61.

[25] S. Chen and D. Donoho, "Basis pursuit", in *Signals, Systems and Computers, Conference Record of the Twenty-Eighth Asilomar Conference on*, vol. 1, pp. 41–44, Oct. 1994.

[26] T. Chen and B. A. Francis, *Optimal Sampled-Data Control Systems.* Springer, 1995.

[27] J. F. Claerbout and F. Muir, "Robust modeling with erratic data", *Geophysics*, vol. 38, no. 5, 1973, pp. 826–844.

[28] F. Clarke, *Functional Analysis, Calculus of Variations and Optimal Control*, vol. 264, ser. Graduate Texts in Mathematics. Springer, London, 2013, pp. xiv+591. DOI: 10.1007/978-1-4471-4820-3.

[29] P. L. Combettes and J.-C. Pesquet, "Proximal splitting methods in signal processing", in *Fixed-Point Algorithms for Inverse Problems in Science and Engineering*, New York, NY: Springer New York, 2011, pp. 185–212.

[30] T. H. Cormen, C. E. Leiserson, R. L. Rivest, and C. Stein, *Introduction to Algorithms*, 3rd. MIT Press, 2009.

[31] T. M. Cover and J. A. Thomas, *Elements of Information Theory*, 2nd. Wiley–Interscience, 2006.

[32] M. S. Darup, G. Book, D. E. Quevedo, and M. Nagahara, "Fast hands-off control using admm real-time iterations", *IEEE Trans. Autom. Control*, vol. 67, no. 10, 2021, pp. 5416–5423.

[33] G. M. Davis, S. G. Mallat, and Z. Zhang, "Adaptive time-frequency decompositions", *Optical Engineering*, vol. 33, no. 7, 1994, pp. 2183–2191.

[34] N. K. Dhingra, M. R. Jovanović, and Z. Luo, "An ADMM algorithm for optimal sensor and actuator selection", in *53rd IEEE Conference on Decision and Control*, pp. 4039–4044, 2014.

[35] R. Doelman and M. Verhaegen, "Sequential convex relaxation for robust static output feedback structured control", in *IFAC-PapersOnLine*, pp. 15 518–15 523, 2017.

[36] D. L. Donoho, "Compressed sensing", *IEEE Trans. Inf. Theory*, vol. 52, no. 4, Apr. 2006, pp. 1289–1306.

[37] D. L. Donoho and P. B. Stark, "Uncertainty principles and signal recovery", *SIAM Journal on Applied Mathematics*, vol. 49, no. 3, 1989, pp. 906–931.

[38] R. Dorfman, "The detection of defective members of large populations", *Ann. Math. Statist.*, vol. 14, no. 4, Dec. 1943, pp. 436–440.

[39] G.-R. Duan and H.-H. Yu, *LMIs in Control Systems.* CRC Press, 2013.

[40] B. Dunham, "Automatic on/off switching gives 10-percent gas saving", *Popular Science*, vol. 205, no. 4, Oct. 1974, p. 170.

[41] J. Eckstein and D. Bertsekas, "On the Douglas-Rachford splitting method and proximal point algorithm for maximal monotone operators", *Math. Program.*, vol. 55, 1992, pp. 293–318.

[42] M. B. Egerstedt and C. F. Martin, *Control Theoretic Splines: Optimal Control, Statistics, and Path Planning.* Princeton University Press, 2009.

[43] M. Elad, *Sparse and Redundant Representations.* Springer, 2010.

[44] M. Fazel, H. Hindi, and S. Boyd, "Rank minimization and applications in system theory", in *Proceedings of the 2004 American Control Conference*, vol. 4, pp. 3273–3278, 2004.

[45] S. Foucart and H. Rauhut, *A Mathematical Introduction to Compressive Sensing.* Birkhäuser, 2013.

[46] M. Gallieri and J. M. Maciejowski, "ℓ_{asso} MPC: Smart regulation of over-actuated systems", in *Proc. Amer. Contr. Conf.* Jun. 2012, pp. 1217–1222.

[47] C. Giraud, *Introduction to High-Dimensional Statistics.* CRC Press, 2015.

[48] G. H. Golub and C. F. V. Loan, *Matrix Computations*, 4th. Johns Hopkins University Press, 2012.

[49] I. Goodfellow, Y. Bengio, and A. Courville, *Deep Learning.* MIT Press, 2016.

[50] D. A. Harville, *Matrix Algebra From a Statistician's Perspective.* Springer, 1997.

[51] T. Hastie, R. Tibshirani, and J. Friedman, *The Elements of Statistical Learning.* Springer, 2009.

[52] T. Hastie, R. Tibshirani, and M. Wainwright, *Statistical Learning with Sparsity: The Lasso and Generalizations.* CRC Press, 2015.

[53] R. Hayakawa and K. Hayashi, "Discreteness-aware approximate message passing for discrete-valued vector reconstruction", *IEEE Trans. Signal Process.*, vol. 66, no. 24, 2018, pp. 6443–6457.

[54] K. Hayashi, M. Nagahara, and T. Tanaka, "A user's guide to compressed sensing for communications systems", *IEICE Trans. on Communications*, vol. E96-B, no. 3, Mar. 2013, pp. 685–712.

[55] N. Hayashi, T. Ikeda, and M. Nagahara, "Design of sparse control with minimax concave penalty", *IEEE Control Systems Letters*, 2024.

[56] W. P. M. H. Heemels, K. H. Johansson, and P. Tabuada, "An introduction to event-triggered and self-triggered control", in *2012 IEEE 51st IEEE Conference on Decision and Control (CDC)*, pp. 3270–3285, Dec. 2012.

[57] H. Hermes and J. P. Lasalle, *Functional Analysis and Time Optimal Control.* Academic Press, 1969.

[58] T. Ikeda and K. Kashima, "Sparsity-constrained controllability maximization with application to time-varying control node selection", *IEEE Control Systems Letters*, vol. 2, 2018, pp. 321–326.

[59] T. Ikeda and K. Kashima, "On sparse optimal control for general linear systems", *IEEE Trans. Autom. Control*, vol. 64, no. 5, 2019, pp. 2077–2083.

[60] T. Ikeda, M. Nagahara, and K. Kashima, "Maximum hands-off distributed control for consensus of multi-agent systems with sampled-data state observation", *IEEE Trans. Control Netw. Syst.*, vol. 6, no. 2, Jun. 2019, pp. 852–862.

[61] T. Ikeda, M. Nagahara, and S. Ono, "Discrete-valued control of linear time-invariant systems by sum-of-absolute-values optimization", *IEEE Trans. Autom. Control*, vol. 62, no. 6, 2017, pp. 2750–2763.

[62] T. Ikeda, D. Zelazo, and K. Kashima, "Maximum hands-off distributed bearing-based formation control", in *2019 IEEE 58th Conference on Decision and Control (CDC)*, pp. 4459–4464, 2019.

[63] T. Ikeda, "Non-convex optimization problems for maximum hands-off control", *IEEE Trans. Autom. Control*, 2024, pp. 1–8.

[64] T. Ikeda and M. Nagahara, "Value function in maximum hands-off control for linear systems", *Automatica*, vol. 64, 2016, pp. 190–195.

[65] T. Ikeda and M. Nagahara, "Discrete-valued model predictive control using sum-of-absolute-values optimization", *Asian Journal of Control*, vol. 20, no. 1, 2018, pp. 196–206.

[66] T. Ikeda and M. Nagahara, "Time-optimal hands-off control for linear time-invariant systems", *Automatica*, vol. 99, 2019, pp. 54–58.

[67] T. Ikeda and M. Nagahara, "Maximum hands-off control with time-space sparsity", *IEEE Contr. Syst. Lett.*, vol. 5, no. 4, 2020, pp. 1213–1218.

[68] T. Ikeda and M. Nagahara, "Resource-aware time-optimal control with multiple sparsity measures", *Automatica*, vol. 135, 2022, p. 109 957.

[69] M. Ishikawa, "Structural learning with forgetting", *Neural Netw.*, vol. 9, no. 3, Apr. 1996, pp. 509–521.

[70] A. Jadbabaie, A. Olshevsky, G. J. Pappas, and V. Tzoumas, "Minimal reachability is hard to approximate", *IEEE Trans. Autom. Control*, vol. 64, no. 2, 2019, pp. 783–789.

[71] D. Jeong and W. Jeon, "Performance of adaptive sleep period control for wireless communications systems", *IEEE Trans. Wireless Commun.*, vol. 5, no. 11, Nov. 2006, pp. 3012–3016.

[72] G. Joseph, "Sparse actuator control of discrete-time linear dynamical systems", *Foundations and Trends in Systems and Control*, vol. 11, no. 3, 2024, pp. 186–284.

[73] G. Joseph and C. R. Murthy, *Sparsity-Constrained Linear Dynamical Systems.* Springer, 2024.

[74] M. R. Jovanović and N. K. Dhingra, "Controller architectures: Tradeoffs between performance and structure", *European Journal of Control*, vol. 30, 2016, pp. 76–91.

[75] N. Karumanchi, *Data Structures and Algorithms Made Easy*, 2nd. CareerMonk, 2011.

[76] E. Khmelnitsky, "On an optimal control problem of train operation", *IEEE Trans. Autom. Control*, vol. 45, no. 7, 2000, pp. 1257–1266.

[77] R. Kirchhoff, M. Thele, M. Finkbohner, P. Rigley, and W. Settgast, "Start-stop system distributed in-car intelligence", *ATZextra worldwide*, vol. 15, no. 11, Jan. 2010, pp. 52–55.

[78] M. Kishida and M. Nagahara, "Risk-aware maximum hands-off control using worst-case conditional value-at-risk", *IEEE Trans. Autom. Control*, vol. 68, no. 10, 2023, pp. 6353–6360.

[79] E. Kreyszig, *Introductory Functional Analysis with Applications.* Wiley, 1989.

[80] Y. Kumar, S. Srikant, D. Chatterjee, and M. Nagaraha, "Sparse optimal control problems with intermediate constraints: Necessary conditions", *Optimal Control Applications and Methods*, vol. 43, no. 2, 2022, pp. 369–385.

[81] F. Leibfritz, *Compl$_e$ib: Constraint matrix optimization problem library*, version Version 1.1, 2005. URL: http://www.complib.de/.

[82] D. Liberzon, *Calculus of Variations and Optimal Control Theory: A Concise Introduction.* Princeton University Press, 2012.

[83] F. Lin, M. Fardad, and M. R. Jovanović, "Augmented Lagrangian approach to design of structured optimal state feedback gains", *IEEE Trans. Autom. Control*, vol. 56, no. 12, 2011, pp. 2923–2929.
[84] F. Lin, M. Fardad, and M. R. Jovanović, "Design of optimal sparse feedback gains via the alternating direction method of multipliers", *IEEE Trans. Autom. Control*, vol. 58, no. 9, 2013, pp. 2426–2431.
[85] L. Ljung, *System Identification: Theory for the User*, 2nd. Upper Saddle River, New Jersey: Prentice Hall, 1999.
[86] B. F. Logan, "Properties of high-pass signals", Ph.D. dissertation, Columbia University, 1965.
[87] C. Louizos, M. Welling, and D. P. Kingma, "Learning sparse neural networks through L_0 regularization", *arXiv preprint arXiv:1712.01312*, 2017.
[88] J. M. Maciejowski, *Predictive Control with Constraints*. Prentice-Hall, 2002.
[89] S. G. Mallat and Z. Zhang, "Matching pursuits with time-frequency dictionaries", *IEEE Trans. Signal Process.*, vol. 41, no. 12, Nov. 1993, pp. 3397–3415.
[90] S. Mallat, *A Wavelet Tour of Signal Processing: The Sparse Way*, 3rd. Academic Press, 2008.
[91] I. Markovsky, *Low Rank Approximation*. Springer, 2012.
[92] R. Martin, K. L. Teo, and M. D'lncalci, *Optimal Control of Drug Administration in Cancer Chemotherapy*. Singapore: World Scientific, 1994.
[93] M. Mesbahi and G. P. Papavassilopoulos, "On the rank minimization problem over a positive semidefinite linear matrix inequality", *IEEE Trans. Autom. Control*, vol. 42, no. 2, Feb. 1997, pp. 239–243.
[94] D. K. Molzahn, F. Dörfler, H. Sandberg, S. H. Low, S. Chakrabarti, R. Baldick, and J. Lavaei, "A survey of distributed optimization and control algorithms for electric power systems", *IEEE Trans. Smart Grid*, vol. 8, no. 6, 2017, pp. 2941–2962.
[95] K. Motonaka, T. Watanabe, Y. Kwon, M. Nagahara, and S. Miyoshi, "Control of a quadrotor group based on maximum hands-off distributed control", *International Journal of Mechatronics and Automation*, vol. 8, no. 4, 2021, pp. 200–207.
[96] U. Münz, M. Pfister, and P. Wolfrum, "Sensor and actuator placement for linear systems based on H_2 and H_∞ optimization", *IEEE Trans. Autom. Control*, vol. 59, no. 11, 2014, pp. 2984–2989.
[97] M. Nagahara, "Discrete signal reconstruction by sum of absolute values", *IEEE Signal Process. Lett.*, vol. 22, no. 10, Oct. 2015, pp. 1575–1579.
[98] M. Nagahara, S. Azuma, and H. Ahn, *Control of Multi-agent Systems—Theory and Simulations with Python*. Springer, 2024.
[99] M. Nagahara and C. F. Martin, "Monotone smoothing splines using general linear systems", *Asian Journal of Control*, vol. 5, no. 2, Mar. 2013, pp. 461–468.
[100] M. Nagahara and C. F. Martin, "L^1 control theoretic smoothing splines", *IEEE Signal Process. Lett.*, vol. 21, no. 11, Nov. 2014, pp. 1394–1397.
[101] M. Nagahara, T. Matsuda, and K. Hayashi, "Compressive sampling for remote control systems", *IEICE Trans. on Fundamentals*, vol. E95-A, no. 4, Apr. 2012, pp. 713–722.
[102] M. Nagahara, M. Ogura, and Y. Yamamoto, "Iterative greedy LMI for sparse control", *IEEE Contr. Syst. Lett.*, vol. 6, 2022, pp. 986–991.
[103] M. Nagahara and D. E. Quevedo, "Sparse representations for packetized predictive networked control", in *IFAC 18th World Congress*, pp. 84–89, Aug. 2011.

[104] M. Nagahara, D. E. Quevedo, and D. Nešić, "Maximum hands-off control and L^1 optimality", in *52nd IEEE Conference on Decision and Control (CDC)*, pp. 3825–3830, Dec. 2013.

[105] M. Nagahara, D. E. Quevedo, and D. Nešić, "Maximum hands-off control: A paradigm of control effort minimization", *IEEE Trans. Autom. Control*, vol. 61, no. 3, 2016, pp. 735–747.

[106] M. Nagahara, D. Quevedo, and J. Østergaard, "Sparse packetized predictive control for networked control over erasure channels", *IEEE Trans. Autom. Control*, vol. 59, no. 7, Jul. 2014, pp. 1899–1905.

[107] M. Nagahara, "Sparse control for continuous-time systems", *International Journal of Robust and Nonlinear Control*, vol. 33, no. 1, 2023, pp. 6–22.

[108] M. Nagahara, "Introduction to compressed sensing with python", *IEICE Transactions on Communications*, vol. 107, no. 1, 2024, pp. 126–138.

[109] M. Nagahara, S.-I. Azuma, and H.-S. Ahn, *Control of Multi-agent Systems: Theory and Simulations with Python.* Springer Nature, 2024.

[110] M. Nagahara, D. Chatterjee, N. Challapalli, and M. Vidyasagar, "CLOT norm minimization for continuous hands-off control", *Automatica*, vol. 113, 2020, p. 108 679.

[111] M. Nagahara, Y. Fujimoto, and Y. Yamamoto, "Sparse system identification with kernel regularization", in *25th International Symposium on Mathematical Theory of Networks and Systems (MTNS2022)*, 2022.

[112] M. Nagahara, Y. Iwai, and N. Sebe, "Projection onto the set of rank-constrained structured matrices for reduced-order controller design", *Algorithms*, vol. 15, no. 9, 2022, p. 322.

[113] M. Nagahara, J. Østergaard, and D. E. Quevedo, "Discrete-time hands-off control by sparse optimization", *EURASIP Journal on Advances in Signal Processing*, vol. 2016, no. 1, 2016, pp. 1–8.

[114] M. Nagahara and Y. Yamamoto, "A survey on compressed sensing approach to systems and control", *Mathematics of Control, Signals, and Systems*, vol. 36, no. 1, 2024, pp. 1–20.

[115] M. Nalbach, A. Korner, and S. Kahnt, "Active engine-off coasting using 48V: Economic reduction of CO_2 emissions", in *17th International Congress ELIV*, pp. 41–51, Oct. 2015.

[116] A. Nedic, "Distributed gradient methods for convex machine learning problems in networks: Distributed optimization", *IEEE Signal Processing Magazine*, vol. 37, no. 3, 2020, pp. 92–101.

[117] A. Nedic and A. Ozdaglar, "Distributed subgradient methods for multi-agent optimization", *IEEE Transactions on Automatic Control*, vol. 54, no. 1, 2009, pp. 48–61.

[118] D. Needell and J. A. Tropp, "CoSaMP: Iterative signal recovery from incomplete and inaccurate samples", *Appl. Comput. Harmonic Anal.*, vol. 26, no. 3, 2008, pp. 301–321.

[119] K. Ogata, *Modern Control Engineering*, 5th. Pearson, 2009.

[120] A. Olshevsky, "Minimal controllability problems", *IEEE Trans. Control Netw. Syst.*, vol. 1, no. 3, 2014, pp. 249–258.

[121] S. K. Pakazad, H. Ohlsson, and L. Ljung, "Sparse control using sum-of-norms regularized model predictive control", in *52nd IEEE Conference on Decision and Control*, pp. 5758–5763, 2013.

[122] N. Parikh and S. Boyd, "Proximal algorithms", *Foundations and Trends in Optimization*, vol. 1, no. 3, 2013, pp. 123–231.

[123] F. Pasqualetti, S. Zampieri, and F. Bullo, "Controllability metrics, limitations and algorithms for complex networks", *IEEE Trans. Control Netw. Syst.*, vol. 1, no. 1, 2014, pp. 40–52.

[124] Y. C. Pati, R. Rezaiifar, and P. S. Krishnaprasad, "Orthogonal matching pursuit: Recursive function approximation with applications to wavelet decomposition", in *Proc. the 27th Annual Asilomar Conf. on Signals, Systems and Computers*, pp. 40–44, Nov. 1993.

[125] S. Pequito, S. Kar, and A. P. Aguiar, "A framework for structural input/output and control configuration selection of large-scale systems", *IEEE Trans. Autom. Control*, vol. 61, no. 2, Feb. 2016, pp. 303–318.

[126] G. Pillonetto, T. Chen, A. Chiuso, G. De Nicolao, and L. Ljung, *Regularized system identification: Learning dynamic models from data.* Springer Nature, 2022.

[127] B. Polyak, M. Khlebnikov, and P. Shcherbakov, "An LMI approach to structured sparse feedback design in linear control systems", in *2013 European Control Conference (ECC)*, pp. 833–838, 2013.

[128] L. S. Pontryagin, *Mathematical Theory of Optimal Processes*, vol. 4. CRC Press, 1987.

[129] B. Recht, M. Fazel, and P. A. Parrilo, "Guaranteed minimum-rank solutions of linear matrix equations via nulcear norm minimization", *SIAM Review*, vol. 52, no. 3, 2010, pp. 451–501.

[130] B. Recht, W. Xu, and B. Hassibi, "Necessary and sufficient conditions for success of the nuclear norm heuristic for rank minimization", in *2008 47th IEEE Conference on Decision and Control*, pp. 3065–3070, 2008.

[131] L. I. Rudin, S. Osher, and E. Fatemi, "Nonlinear total variation based noise removal algorithms", *Physica D*, vol. 60, 1992, pp. 259–268.

[132] W. Rudin, *Principles of Mathematical Analysis*, 3rd International. McGraw-Hill, 1976.

[133] W. Rudin, *Real and Complex Analysis*, 3rd International. McGraw-Hill, 2005.

[134] F. Santosa and W. W. Symes, "Linear inversion of band-limited reflection seismograms", *SIAM Journal on Scientific and Statistical Computing*, vol. 7, no. 4, 1986, pp. 1307–1330.

[135] H. Sasahara, K. Hayashi, and M. Nagahara, "Symbol detection for faster-than-Nyquist signaling by sum-of-absolute-values optimization", *IEEE Signal Process. Lett.*, vol. 23, no. 12, 2016, pp. 1853–1857.

[136] H. Sasahara, K. Hayashi, and M. Nagahara, "Multiuser detection based on MAP estimation with sum-of-absolute-values relaxation", *IEEE Trans. Signal Process.*, vol. 65, no. 21, 2017, pp. 5621–5634.

[137] S. Scardapane, D. Comminiello, A. Hussain, and A. Uncini, "Group sparse regularization for deep neural networks", *Neurocomputing*, vol. 241, 2017, pp. 81–89.

[138] H. Schättler and U. Ledzewicz, *Geometric Optimal Control.* Springer, 2012.

[139] B. Schölkopf and A. J. Smola, *Learning with Kernels.* The MIT Press, 2002.

[140] P. Shakouri, A. Ordys, P. Darnell, and P. Kavanagh, "Fuel efficiency by coasting in the vehicle", *International Journal of Vehicular Technology*, vol. 2013, 2013, p. 14.

[141] R. E. Skelton, T. Iwasaki, and K. Grigoriadis, *A Unified Algebraic Approach to Linear Control Design.* London: Taylor & Francis, 1998.

[142] E. D. Sontag, *Mathematical Control Theory*, second. New York: Springer, 1998.

[143] N. Srivastava, G. Hinton, A. Krizhevsky, I. Sutskever, and R. Salakhutdinov, "Dropout: A simple way to prevent neural networks from overfitting", *Journal of Machine Learning Research*, vol. 15, 2014, pp. 1929–1958.

[144] G. Strang and T. Nguyen, *Wavelets and Filter Banks*, 2nd. Wellesley-Cambridge Press, 1996.

[145] S. Sun, M. B. Egerstedt, and C. F. Martin, "Control theoretic smoothing splines", *IEEE Trans. Autom. Control*, vol. 45, no. 12, Dec. 2000, pp. 2271–2279.

[146] H. L. Taylor, S. C. Banks, and J. F. McCoy, "Deconvolution with the ℓ_1 norm", *Geophysics*, vol. 44, no. 1, 1979, pp. 39–52.

[147] R. Tibshirani, "Regression shrinkage and selection via the LASSO", *J. R. Statist. Soc. Ser. B*, vol. 58, no. 1, 1996, pp. 267–288.

[148] V. Tzoumas, M. A. Rahimian, G. J. Pappas, and A. Jadbabaie, "Minimal actuator placement with bounds on control effort", *IEEE Trans. Control Netw. Syst.*, vol. 3, no. 1, 2016, pp. 67–78.

[149] M. Vidyasagar, *An Introduction to Compressed Sensing.* SIAM, 2019.

[150] G. Vossen and H. Maurer, "On L^1-minimization in optimal control and applications to robotics", *Optimal Control Applications and Methods*, vol. 27, no. 6, 2006, pp. 301–321.

[151] W. Wen, C. Wu, Y. Wang, Y. Chen, and H. Li, "Learning structured sparsity in deep neural networks", *Advances in neural information processing systems*, vol. 29, 2016.

[152] Y. Yamamoto, *From Vector Spaces to Function Spaces: Introduction to Functional Analysis with Applications.* SIAM, 2012.

[153] N. Young, *An Introduction to Hilbert Space.* Cambridge University Press, 1988.

[154] K. Yuan, Q. Ling, and W. Yin, "On the convergence of decentralized gradient descent", *SIAM Journal on Optimization*, vol. 26, no. 3, 2016, pp. 1835–1854.

[155] M. Yuan and Y. Lin, "Model selection and estimation in regression with grouped variables", *Journal of the Royal Statistical Society: Series B (Statistical Methodology)*, vol. 68, no. 1, Feb. 2006, pp. 49–67.

[156] K. Zhou, J. C. Doyle, and K. Glover, *Robust and Optimal Control.* Pearson, 1995.

[157] M. Zibulevsky and M. Elad, "L1-L2 optimization in signal and image processing", *IEEE Signal Process. Mag.*, vol. 27, May 2010, pp. 76–88.

[158] H. Zou and T. Hastie, "Regularization and variable selection via the elastic net", *Journal of the Royal Statistical Society: Series B (Statistical Methodology)*, vol. 67, no. 2, Apr. 2005, pp. 301–320.

Index

About the Author

Masaaki Nagahara received a bachelor's degree in engineering from Kobe University in 1998 and a master's degree and a doctoral degree in informatics from Kyoto University in 2000 and 2003, respectively.

He is currently a Full Professor at the Graduate School of Advanced Science and Engineering, Hiroshima University. He has been a Visiting Professor at Indian Institute of Technology Bombay since 2017. His research interests include control theory, machine learning, and sparse modeling.

He received remarkable international awards: Transition to Practice Award in 2012 and George S. Axelby Outstanding Paper Award in 2018 from the IEEE Control Systems Society. Also, he received many awards from Japanese research societies, such as SICE Young Authors Award in 1999, SICE Best Paper Award in 2012, SICE Best Book Authors Awards in 2016 and 2021, SICE Control Division Research Award (Kimura Award) in 2020, and the Best Tutorial Paper Award from the IEICE Communications Society in 2014.

He is a senior member of IEEE, and a member of IEICE, SICE, ISCIE, and RSJ.

www.ingramcontent.com/pod-product-compliance
Lightning Source LLC
LaVergne TN
LVHW051051220926
848485LV00001BA/73